Introduction

The Primary Mathematics curriculum allows students to develop their ability in mathematical problem solving. This includes using and applying mathematics in practical, real-life situations as well as within the discipline of mathematics itself. The curriculum covers a wide range of situations from routine problems to problems in unfamiliar contexts to open-ended investigations that make use of relevant mathematical concepts.

An important feature of learning mathematics with this curriculum is the use of a concrete introduction to the concept, followed by a pictorial representation, followed by the abstract symbols. This occurs both at the level of one lesson and at the level of major themes throughout the curriculum. The textbook supplies the pictorial and abstract aspects of this progression within a lesson. You, as the teacher, should supply the concrete introduction. If the topic is new to the student, provide the concrete introduction before doing the tasks in the text. Use the tasks in the textbook more as an assessment of understanding. If the topic is mostly review, you can use concrete objects if you detect some misunderstanding but otherwise use the pictorial representation. For some students a concrete illustration is more important than for other students.

The textbook and the workbook are a basis for developing mathematical reasoning skills. Practicing procedures or "math facts" can be easily added through additional worksheets for students who need more practice. As you go through this curriculum, it is important that you gauge your own student's understanding of concepts and need for additional practice and provide it when needed. The textbook is meant be used for discussion, learning, and assessment by the teacher as part of the lesson. Some pages or tasks are meant for discussion that can go beyond the obvious answer. Other tasks are straightforward ones where you student simply supplies the answer. The workbook is for independent practice.

For review or reinforcement you can use the supplemental books for Extra Practice and Tests. In the test book, there are two tests for each section. The second test is multiple choice. There is also a set of cumulative tests at the end of each unit. You do not need to use both tests. If you use only one test, you can save the other for review or practice later on.

The Mental Math pages in the appendix can be used at any time after the lesson that refers to them. They can be used multiple times.

The purpose of this guide is to help you to understand the important concepts of the *Primary Mathematics* curriculum, to gain an understanding of how these concepts fit in with the program as a whole, to provide suggestions to help you introduce concepts concretely and use the textbook effectively, and to provide suggestions for additional activities for reinforcement and practice. You should pick and choose the activities that are most useful for your particular student – you do not have to do every activity.

This guide will give you a suggested amount of time in weeks to spend on each unit to help you keep on track for finishing in about 18 weeks. For some units, your child may be able to do the work more quickly, and for others more slowly. Take the time your student needs on each topic, not the time indicated by an arbitrary schedule.

Scheme of Work

Textbook: Primary Mathematics, Standards Edition, 1A Textbook
Workbook: Primary Mathematics, Standards Edition, 1A Workbook
Guide: Primary Mathematics 1A, Standards Edition, Home Instructor's Guide (this book)
Extra Practice: Primary Mathematics, Standards Edition, 1
Tests: Primary Mathematics, Standards Edition, 1A Tests

Week		Objectives	Text book	Work book	Guide
Unit 1: Numbers 0 to 10					
		Chapter 1: Counting			1
1	1	▪ Count to 10, read and write numerals and number words. ▪ Recognize 0 as an empty set.	8-13	7-10	2-3
		Extra Practice, Unit 1, Exercise 1A, pp. 3-4			
	2	▪ Compare numbers within 10. ▪ Understand more and less.	14-15	11-12	4-5
	3	▪ Count on from some number to 10. ▪ Count backward from 10 to 0. ▪ Find one more or one less than a number within 10. ▪ Arrange the numbers 1-10 in order. ▪ Determine missing numbers in a sequence.	16-17	13-14	6
		Extra Practice, Unit 1, Exercise 1B, pp. 5-6			
9/15		*Tests*, Unit 1, 1A and 1B, pp. 1-6			
Unit 2: Number Bonds					
		Chapter 1: Making Number Stories			7-9
2	1	▪ Make up number stories to illustrate number bonds. ▪ Associate number bonds with part-whole. ▪ Divide groups up in different ways. ▪ Find and memorize number pairs that make 2, 3, 4, 5, and 6.	18-20	15	10-11
	2	▪ Find number pairs that make 7. ▪ Find number pairs that make 8. ▪ Find number pairs that make 9. ▪ Find number pairs that make 10.	21-23	16-19	12-14
3	3	▪ Find the missing part of a number bond.	24	20-22	15-16
	4	▪ Review ways to make 10.	25	23-24	17-18
		Extra Practice, Unit 2, Exercise 1, pp. 9-12			
		Tests, Unit 2, 1A and 1B, pp. 7-10			
		Tests, Cumulative Test Units 1-2, A and B, pp. 11-16			

Week		Objectives	Text book	Work book	Guide
Unit 3: Addition					
		Chapter 1: Making Addition Stories			19
	1	▪ Understand the meaning of addition. ▪ Make number stories for addition.	26	25-27	20-21
		Extra Practice, Unit 3, Exercise 1, pp. 15-16			
4	2	▪ Write addition equations using + and =.	27-31	28-30	22-23
		Tests, Unit 3, 1A and 1B, pp. 17-20			
		Chapter 2: Addition with Number Bonds			24-25
	1	▪ Relate addition stories to number bonds. ▪ Write two addition facts for a given number bond.	32-33	31-35	26
	2	▪ Solve picture problems using addition. ▪ Memorize addition facts through 5, +0, +1.	34	36-37	27
		Extra Practice, Unit 3, Exercise 2, pp. 17-18			
		Tests, Unit 3, 2A and 2B, pp. 21-24			
		Chapter 3: Other Methods of Addition			28-29
5	1	▪ **Count on** to add 1, 2, or 3 to a number within 10. ▪ Learn facts for +2, +3.	35-37	38-39	30-31
	2	▪ Learn facts for making 10. ▪ Learn facts for 3 + 4, 4 + 5. ▪ Review addition facts through 10.	38-40	40-41	32-33
		Extra Practice, Unit 3, Exercise 3A-3C, pp. 19-26			
		Tests, Unit 3, 3A and 3B, pp. 25-28			
		Tests, Cumulative Test Units 1-3, A and B, pp. 29-36			
Unit 4: Subtraction					
		Chapter 1: Making Subtraction Stories			34
6	1	▪ Understand the meaning of subtraction. ▪ Make number stories for subtraction.	41	42-45	35-36
	2	▪ Write subtraction equations using the symbols − and =. ▪ Write several subtraction equations for a given situation.	42-46	46-48	37
		Extra Practice, Unit 4, Exercise 1A-1B, pp. 29-34			
		Tests, Unit 4, 1A and 1B, pp. 37-40			

Week		Objectives	Text book	Work book	Guide
		Chapter 2: Methods of Subtraction			38-41
	1	• Relate subtraction facts within 10 to the missing part of a number bond. • Write two subtraction facts for a given number bond.	47-49	49-52	42
7	2	• Write two addition equations and two subtraction equations for a given number bond.	50	53-56	43-44
	3	• Count back to subtract 1, 2, or 3 from a number within 10. • Memorize subtraction facts for –0, –1, –2, –3.	51-52	57-58	45-46
	4	• Recognize numbers that differ by 1 or 2. • Count on to subtract numbers close to each other. • Subtract from 10.	53	59	47-48
8	5	• Review addition and subtraction within 10.	54-55	60, 67	49
	6	• Interpret addition and subtraction stories.		61-66	50-51
		Extra Practice, Unit 4, Exercise 2A-2C, pp. 35-42			
		Tests, Unit 4, 2A and 2B, pp. 41-44			
		Tests, Cumulative Test Units 1-4, A and B, pp. 45-51			
Unit 5: Position					
		Chapter 1: Position and Direction			52
9	1	• Use position words. • Use left and right in directions.	56-59	68-70	53
		Extra Practice, Unit 5, Exercise 1, pp. 47-48			
		Tests, Unit 5, 1A and 1B, pp. 53-56			
		Chapter 2: Naming Positions			54
	1	• Name a position using ordinal numbers 1st through 10th. • Find an ordinal position from the left or from the right.	60-61	71-75 64-75	55
		Extra Practice, Unit 5, Exercise 2, pp. 49-50			
		Tests, Unit 5, 2A and 2B, pp. 57-60			
Review					
		• Review all topics.		76-88	56-57
		Tests, Cumulative Test Units 1-5, A and B, pp. 61-68			

Week		Objectives	Text book	Work book	Guide
Unit 6: Numbers to 20					
		Chapter 1: Counting and Comparing			58-59
10	1	▪ Count to 20 by building up from 10. ▪ Read and write numerals and number words for 11 to 19.	62-64	89-93	60-61
	2	▪ Relate numbers 11 to 20 to a ten plus ones.	65-66	94-95	62
	3	▪ Count backwards from 20. ▪ Determine missing numbers in a sequence from 1 to 20.	66-67	96-97	63
	4	▪ Compare and order numbers within 20.	68-69	98-100	64-65
		Extra Practice, Unit 6, Exercise 1, pp. 55-58			
		Tests, Unit 6, 1A and 1B, pp. 69-74			
		Chapter 2: Addition and Subtraction			66-67
11	1	▪ Add 1-digit numbers whose sum is greater than 10 by first making a ten.	70-72	101-104	68-70
	2	▪ Add a 1-digit number to a 2-digit number within 20 (no renaming) by adding ones.	73	105-107	71
		Extra Practice, Unit 6, Exercise 2A, pp. 59-60			
12	3	▪ Subtract a 1-digit number from a 2-digit number within 20 when there are enough ones (no renaming).	73-74	108-110	72-73
	4	▪ Subtract a 1-digit number from a 2-digit number within 20 when there are not enough ones by subtracting from the ten.	74	111-113	74-76
		Extra Practice, Unit 6, Exercise 2B, pp. 61-62			
	5	▪ Add 1, 2, or 3 to a number within 20 by counting on. ▪ Subtract 1, 2, or 3 from a number within 20 by counting back.	75	114-115	77
13	6	▪ Write a family of 2 addition and 2 subtraction equations for a given fact. ▪ Practice basic addition and subtraction facts within 20.	76-78	116-123	78-79
		Extra Practice, Unit 6, Exercise 2C-2D, pp. 63-68			
		Tests, Unit 6, 2A and 2B, pp. 75-81			
Review					
14		▪ Review all topics.		124-131	80
		Tests, Cumulative Test Units, 1-6, A and B, pp. 83-88			

Week		Objectives	Text book	Work book	Guide .
Unit 7: Shapes					
		Chapter 1: Common Shapes			81
15	1	▪ Recognize and name the four basic shapes: circle, triangle, square, and rectangle.	79-83	132-141	82
		Extra Practice, Unit 7, Exercise 1A, pp. 73-76			
	2	▪ Sort and classify objects by shape, size, color, or orientation.	84-86	142-146	83
	3	▪ Describe or continue a pattern according to one or two attributes such as shape, size, or color.	87-88	147-148	84-85
		Extra Practice, Unit 7, Exercise 1B, pp. 77-78			
	4	▪ Choose suitable shapes to fit together to make a basic shape.	89-90	149	86
		Tests, Unit 7, 1A and 1B, pp. 89-96			
		Tests, Cumulative Test Units 1-7, A and B, pp. 97-104			
Unit 8: Length					
		Chapter 1: Comparing Length			87
16	1	▪ Compare the length of two or more objects by direct comparison or by counting units. ▪ Compare the length of two or more objects by indirect comparison. ▪ Arrange objects in order by length.	91-94	150-153	88
		Extra Practice, Unit 8, Exercise 1, pp. 81-82			
		Tests, Unit 8, 1A and 1B, pp. 105-108			
		Chapter 2 : Measuring Length			89
	1	▪ Estimate and measure length with non-standard units.	95-96	154-156	90-91
		Extra Practice, Unit 8, Exercise 2, pp. 83-84			
		Tests, Unit 8, 2A and 2B, pp. 109-112			
Unit 9: Weight					
		Chapter 1: Comparing Weight			92-93
	1	▪ Compare the weight of two or more objects by direct comparison or by counting identical units.	97-99	157-158	94
		Extra Practice, Unit 9, Exercise 1, pp. 87-88			
		Tests, Unit 9, 1A and 1B, pp. 113-116			

Week		Objectives	Text book	Work book	Guide
		Chapter 2 : Measuring Weight			95
17	1	▪ Estimate and measure weight with non-standard units.	100-101	159-162	96
		Extra Practice, Unit 9, Exercise 2, pp. 89-90			
		Tests, Unit 9, 2A and 2B, pp. 117-121			
Unit 10: Capacity					
		Chapter 1: Comparing Capacity			97
	1	▪ Compare the capacity of two or more containers.	102-105	163-165	98
		Extra Practice, Unit 10, Exercise 1, pp. 93-94			
		Tests, Unit 10, 1A and 1B, pp. 123-128			
		Chapter 2 : Measuring Capacity			99
	1	▪ Measure capacity with non-standard units.	106-107	166-167	100
		Extra Practice, Unit 10, Exercise 2, pp. 95-96			
		Tests, Unit 8, 2A and 2B, pp. 129-132			
Review					
18		▪ Review all topics.		168-176	101
		Tests, Cumulative Test Units 1-10, A and B, pp. 133-142			
Answers to Mental Math					102-104
Appendix – Mental Math					a1-a9
Appendix – Manipulatives					a10-a25

Manipulatives

It is important to introduce the concepts concretely, but it is not important exactly what manipulative is used. A few possible manipulatives are suggested here. The linking cubes and number cards will be used at many levels of the *Primary Mathematics*.

Multilink cubes

These are cubes that can be linked together on all 6 sides. Ten of them can be connected to form tens so that you have tens and ones. You can use Lego's™ or anything else that can be grouped into tens, but the multilink cubes will be useful when the students get to volume problems in *Primary Mathematics* 4.

Four sets of cards:

You can use index cards to make them, but use crayon or pencil rather than marker so the numbers don't show through on the back side. You can also copy the cards in the appendix, ideally onto cardstock, and cut them out.

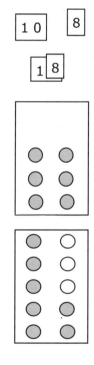

- ➤ Set 1: Numerals 0-20. Include one longer card for 10 so that when teaching numbers from 10 to 19, you can slide the single digits over the 10 card.
- ➤ Set 2: Number words one to twenty
- ➤ Set 3: Dot cards for 0-10. You can use index cards and trace around pennies, or use dot stickers that are available as office supply. Have cards with 1, 2, 3, 4, or 5 down one side for the numbers 1-5, cards with 5 down one side and 1, 2, 3, 4, or 5 down the other side for numbers 6-10. Make some extra ones for 10.
- ➤ Set 4: Dot cards similar to set 3, but cards showing doubles for 2, 3, and 4 (the example here is for double 3), and cards showing doubles + 1 (add another dot to one of the rows showing double 3 to show 3 + 4).
- ➤ Set 5: Dot cards for "making 10". Put 10 empty circles on each card, 5 on each side, and color in 1, 2, 3, 4, 5, 6, 7, 8, or 9 circles, going down one side before the other. (The example here is for 7.)

Counters
Round counters are easy to use and pick up, and can be used for number discs later, but any type of counter will work.

Number cards 0-10
You need four sets of number cards 0-10. You can use a deck of cards and call the ace 1. You can white-out the ace and replace it with a 1. Use the cards to play games that help your student memorize the math facts. You can whiteout the J on the Jack and call that a 0, or use the tens from another deck and whiteout the 1 of the 10 and all the symbols.

Fact cards
A set of addition and subtraction fact cards (for addition and subtraction within 10). These can have the fact on one side and the answer on the other for individual practice.

Fact game cards
Fact cards with answers on separate cards rather than on the back of the card with the addition or subtraction fact.

Whiteboard and dry-erase markers
This guide does not assume you will be teaching in the same way as in a classroom, i.e. with a white board on the wall. Smaller whiteboards that can be used at the table, or by the student to work out problems on, instead of paper, are ideal.

Hundred-chart
Make one or buy one with squares large enough to cover with counters or coins. There is one in the appendix that can be copied.

Number cubes or dice
Some of the games or activities use number cubes. You can get blank cubes with labels, or simply label regular dice using masking tape. Several 10-sided dice are handy for games and activities, but not required.

Simple balance

Supplements

The textbook and workbook provide the kernel of the math curriculum. Some students profit by additional practice, more review. Other students profit by more challenging problems. There are several supplementary workbooks available at www.singaporemath.com. If you feel it is important that your student have a lot of drill in math facts, there are many websites that generate worksheets according to your specifications, or provide on-line fact practice. Web sites come and go, but doing a search using the terms "math fact practice" will turn up many sites. Playing simple card games is another way to practice math facts.

Unit 1 – Numbers 0 to 10

Chapter 1 – Counting

Objectives

- Count to 10, read and write numerals and number words.
- Recognize 0 as an empty set.
- Compare numbers within 10.
- Understand **more** and **less**.
- Count on from some number to 10.
- Count backward from 10 to 0.
- Find one more or one less than a number within 10.
- Arrange the numbers 1-10 in order.
- Determine missing numbers in a sequence.

Material

- Objects to count
- Number cards 0-10
- Number word cards, one to ten
- Dot cards 0-10
- Counters
- Multilink cubes
- Playing cards 0-10

Notes

Before starting *Primary Mathematics* most students have already been counting and using numbers beyond 10 in daily situations, and can also read numerals (number symbols). Some students may also be able to read and write number words. Unit 1 is primarily meant to reinforce these skills. Writing number words is not a necessary skill at this point, and can be further practiced at other times in other contexts. Recognizing (reading) the number words, however, is helpful unless you are reading to your child.

Most students already understand that numbers have order, and each number is one more than the previous number. Number order is reviewed in this unit. Student will also compare sets of objects concretely and pictorially where they can count the objects in each set and match the objects in one set to the objects in the other set to see which set has more. It is not necessary right now to get your student to tell you how many there are more or less in one set than the other. In *Primary Mathematics* 1B, students will learn to use the symbols > and < for greater than and less than.

(1) Review counting, numerals, and number words

Textbook

Pages 8-11

Task 1, pp. 12-13

1. 4	1
0	3
2	5
10	
6	8
9	7

Workbook

Exercise 1, pp. 7-8

1. four   one   three   five
   zero   two

2. nine   seven   five   six
   ten   eight

Exercise 2, pp. 9-10

1. 4	7
10	5
9	6
2	8

2. Check that the correct amounts are colored.

Activity

Ask your student to count sets of objects, such as counters, multilink cubes, toy cars, toy animals, chairs at the table, etc. Make sure she uses one-to-one correspondence; that is, counts one object for one number.

Give your student a set of number cards. Display a set of ten or less objects and ask him to select the correct number card.

Set out number cards and dot cards at random and ask your student to match them.

Put the number word cards in order and read them out loud to your student. Then ask her to read them along with you, then alone. Then mix them up and ask her to read them out loud.

Set out numeral cards and number word cards and ask your student to match them.

Discussion

Discuss pages 8-11 in the textbook. Point out the different ways to show the number of items; an equivalent number of dots, a numeral, a number word. In particular, point out the empty set and make sure your student understands that the numeral 0 represents none or no objects.

You can have your student write the numeral, number word, and show the numbers using dot cards.

Practice

Task 1, pp. 12-13

Workbook

Exercises 1-2, pp. 7-8

Reinforcement

Show numeral cards at random and ask your student to count out the correct number of counters.

Show numeral cards at random and ask your student to write the number words.

Show your student a set of objects and see if he can guess how many there are without counting.

Show number word cards at random and ask your student to say or write the numeral, or count out the correct number of counters.

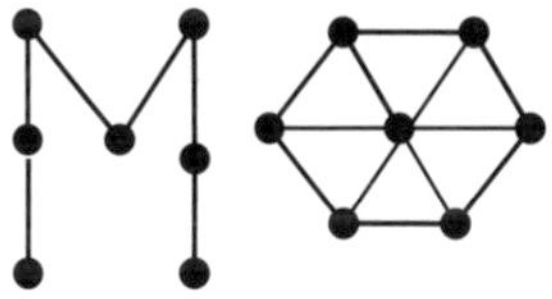

Ask your student to make different dot patterns for a given number of dots. Some dot patterns for 7 are shown here.

Extra Practice, Unit 1, Exercises 1A, pp. 3-4

Game: Go Fish variation

Material: Numeral cards, number word cards, and dot cards. (This gives you a deck of 30 cards, including 0's.)

Goal: To make sets of all three cards for a number.

Procedure: Mix up the cards and deal out 5 cards to each player. Place the remaining cards face down in the center. The first player asks another player if he or she has a number that the first player is holding. For example, the player has a dot card for 6. He asks another player for all her sixes. If she has any sixes, she must give them all to him, and he can ask her or another player for cards. If she does not, she says "Go fish" (or other term), and he draws a card from the pile in the center. Then it is the next player's turn. When a player has a set of three cards, he or she puts them down face up.

(2) Review more and less, comparing numbers

Textbook

Tasks 2-5, pp. 14-15

2. (a) yes
   (b) no
   (c) no

3. 6

4. The dolls

5. The butterflies

Workbook

Exercise 3, pp. 11-12

1. (a) 2nd and 4th
   (b) 1st and 4th
   (c) 2nd and 4th
   (d) 1st and 3rd
   (e) 1st and 4th

2. (a) glasses
   (b) mice

3. (a) flowers
   (b) heads

Activity

Use two types of objects, such as two colors of counters or multilink cubes, or two types of toys. Display two sets, one of each type of object with the same number of objects in each set, such as four objects.

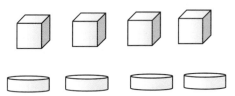

Ask your student to count the number in each set. Ask him if the numbers are the same. They are.

Add one item to one of the sets.

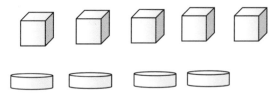

Ask your student how many are in each set. Then ask her if both sets have the same number. They do not.

Ask which set has **more** and which has **less**. Lead your student to say:

➢ There are more cars than planes.
➢ 5 is more than 4.
➢ There are fewer planes than cars.
➢ 4 is less than 5.

Display two sets of objects, set out randomly. Ask your student to decide which set has more or which set has less. Let him line them up.

Repeat the above activity, but without letting your student move the objects around to line them up. If necessary, you can either place the objects on paper so that she can draw lines between pairs of objects, or you can ask her to count the number of objects in each set.

Practice

Tasks 2-5, pp. 14-15

Workbook

Exercise 3, pp. 11-12

Reinforcement

Use playing cards 1- 10. Shuffle the cards and lay down the top two cards face up. Ask your student to tell you which number is greater.

Use numeral cards 1-10. Lay down pairs of numeral cards and ask your student which number is greater. This is more challenging than the previous activity, because the student can't look at objects to compare them.

Game

Material: Playing cards 1-10. In order to keep the game short, you can use half a deck (only two suits).

Goal: To get the most cards at the end.

Procedure: Shuffle the cards and deal out all of them, face down. Players simultaneously turn over one card. The player with the largest number gets all the cards.

(3) Review number order

Textbook

Tasks 6-10, pp. 16-17

7. 10

8. 5

10. (a) 7, 8; 10
    (b) 6; 4, 3
    (c) 7   7
        2   4
        8   0

Workbook

Exercise 4, pp. 13-14

2. (from starting position
   at bottom of page)
   3
   10
   8
   7
   0
   5

Activity

Ask your student to make stacks of 1, 2, 3, 4, 5, 6, 7, 8, 9, and 10 multilink cubes, line them up in order, and place number cards under them (as in task 6, textbook p. 16). Lead him to see that each number is one more than the one before.

Remove the cubes, and then remove several of the numeral cards and mix the removed cards up. Ask your student to replace them in the correct places.

Repeat these activities, but in reverse order, 10 down to 1. Ask your student to practice counting backwards.

Practice

Tasks 6-10, pp. 16-17

Workbook

Exercise 4, pp. 13-14

Reinforcement

Say a number between 1 and 10 and ask your student to count on from that number to 10.

Say a number between 1 and 10 and ask your student to count down from that number to 0.

Mix up a set of number cards 1-10 and ask your student to put them in order, either forward or backward.

Mix up a set of number cards 1-10 and remove 4 of them. Place the remaining cards on the table face up and the removed cards face down. Ask your student which cards are missing, without allowing her to move the cards around.

Give your student 3 to 5 random number cards (not necessarily consecutive) and ask him to put them in order.

Repeat any of these activities with number word cards.

Extra Practice, Unit 1, Exercises 1A, pp. 3-4

Tests

Tests, Unit 1, 1A and 1B, pp. 1-6

Unit 2 – Number Bonds

Chapter 1 – Making Number Stories

Objectives

- Make up number stories to illustrate **number bonds**.
- Associate number bonds with part-whole.
- Find and memorize number pairs for numbers from 2-10
- Find the missing part of a number bond.
- Review ways to make 10.

Material

- Number cards 0-10
- Counters
- Multilink cubes
- Set of double-nine dominoes (optional)

Notes

Number bonds are a combination of three numbers where two of them are parts and one is the whole, or sum, of the other two. A number bond represents a part-whole relationship between 3 numbers. Each number bond is related to a family of four basic addition and subtraction facts. A student can use number bonds to add and subtract. Later, number bonds can be used to show different ways a number can be split up to facilitate mental math or other operations or concepts.

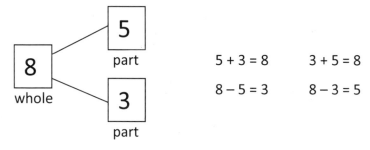

The part-whole concept is important and will be used throughout *Primary Mathematics* along with number bonds. Later students will see that if the whole is missing, the two parts must be added to get the missing whole. If a part is missing, the other part must be subtracted from the whole to get the missing part. This concept will help students determine what equation to use later to solve word problems. So this unit is foundational to the rest of the series.

Your student needs to know the number bonds for 1-10. 1 has only one number bond: 1, 1, 0. There are two or more number bonds for each of the rest.

Make sure your student understands that the orientation of the number bonds does not matter. The whole can be on the top, the bottom, the left side, or the right side of the other two parts. Whether there are circles or squares around the numbers, or nothing around them, also does not matter. What is important is that the drawing distinguishes between the parts and the whole by having lines from the parts to the whole. Thus all of the following are equivalent.

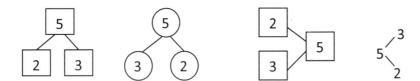

At this stage, it is better to circle or draw squares around the numbers to clearly delineate them from each other. Later, shapes around the numbers will not be needed.

Spend as much time on this material as needed for your student to commit the number bonds for 1-10 to memory. However, if he has a lot of trouble memorizing number bonds, don't stop here long enough to thoroughly bore him; you can continue to have him practice them during the next three units. Some students have an easier time memorizing addition and subtraction facts than number bonds, and some helps will be included in the units on addition and subtraction within 10 to help your student memorize those facts.

Students will use their knowledge of number bonds to find a missing part of a number bond.

The missing part of a number bond can be visualized in two ways.

A set of 8 objects is divided into two parts. One part is 5. What is the other part?

Given a set of 5 objects, how many more are needed to make 8?

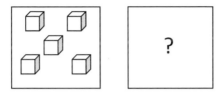

The workbook exercise for this concept includes picture helps. Your student can start out by using objects to find the missing part, but eventually she should be able to do find the missing part without pictures, particularly if she has memorized the number bonds.

The number bonds of 10 are particularly important for addition and subtraction where the total is greater than 10. Give your student plenty of practice in making ten with activities or games.

(1) Number bonds for 2, 3, 4, 5, and 6

Textbook

Pages 18-19

Tasks 1-2, p. 20

1. Answers will vary.

2. 0, 6
   1, 5
   2, 4
   3, 3

Discussion

Text pp. 18-9

You can use the pictures on these pages, or start with toys or other objects that can be grouped in different ways before using the pictures. Ask your student to tell you how many penguins there are and what they are doing. Then ask him to make up number stories for the different pictures. For example:

➢ There are 5 penguins. 2 are swimming. 3 are outside the water.

➢ There are 5 penguins. 2 are big. 3 are small.

➢ There are 5 penguins. 4 are playing. 1 is sleeping.

For each story, point out the number bond, or draw one if using toys or other objects. Lead your student to see that one number is the whole, and the other two numbers are parts. The two parts together make the whole. This is indicated by drawing lines from the part to the whole.

Activity

Ask your student to make a train of 5 multilink cubes. Then ask her to break it apart into two parts in different ways. List all the possible number bonds. Tell her that 0, 5, and 5 also make a number bond. Be sure she understands that splitting 5 into 1 and 4 is the same as splitting 5 into 4 and 1.

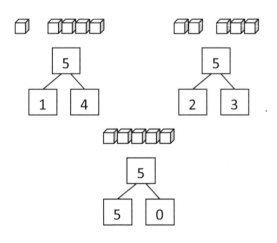

Workbook

Exercise 1, p. 15

1. balloons,
   roller skates,
   bow,
   mittens

Repeat with 6 multilink cubes.

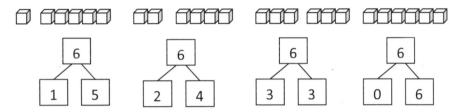

Ask your student to write number bonds for 1, 2, 3, and 4. You can let him use multilink cubes if necessary.

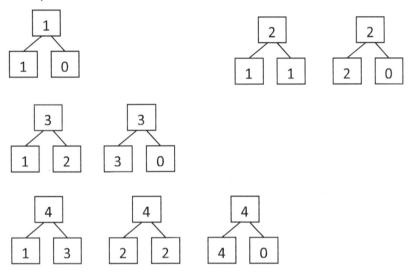

Practice

Tasks 1-2, p. 18

Get your student to come up with more than just one story for task 1. Draw number bonds for each. For example, there are 2 children in the swing and 4 on the seesaw. There are 3 girls and 3 boys. There is 1 child with a cap and 4 children without. You can include number bonds with 0; there are 6 children on the swings or seesaw, and 0 just on the grass.

Workbook

Exercise 1, p. 15

Reinforcement

Use number cards 0-10. Decide on a target number (the whole), such as 5. Set out 5 cards, of which at least two make the target number, such as 2, 3, 4, 6, and 9. Ask your student to pick out the pairs of numbers that make the target number. Repeat with another set of cards, or a different target number.

(2) Find number bonds for 7, 8, 9, and 10

Textbook

Tasks 3-8, pp. 21-23

  3. Answers will vary.

  4. 1, 6; 2, 5; 3, 4; 7, 0

  5. Answers will vary.

  6. 1, 7; 2, 6; 3, 5;
     4, 4; 8, 0

  7. 1, 8; 2, 7; 3, 6;
     4, 5; 9, 0

  8. 1, 9; 2, 8; 3, 7;
     4, 6; 5, 5; 10, 0

Workbook

Exercise 2, p. 16

  1. 1 → 6; 2 → 5 ; 3 → 4;
     4 → 3; 5 → 2; 6 → 3;
     0 → 7

Exercise 3, p. 17

  1. 2, 6   7, 1
     0, 8   3, 5
     4, 4   HELICOPTER

Exercise 4, p. 18

  1. 4 → 5; 5 → 4 ; 6 → 3;
     2 → 7; 7 → 2; 8 → 1
     CIRCUS

Exercise 5, p. 19

  1. 4 → 6; 7 → 3; 8 → 2;
     5 → 5; 0 → 10; 9 → 1

Activity

Ask your student to use two colors of multilink cubes to make all the number bonds for a particular number (7, 8, 9, or 10), such as 1 red and 7 yellows, 2 reds and 6 yellows, 3 reds and 5 yellows, and 4 reds and 4 yellows for 8. She can continue with the pairs that make the same number bonds but using opposite colors, e.g. 5 reds and 3 yellows, 6 reds and 2 yellows, 7 reds and 1 yellow. Point out that these are the same number bonds as found earlier. Write all the number bonds, and include the ones with 0 as a part. The number bonds for 7, 8, 9, and 10 are given on the next page of this guide.

You may wish to spend one day on number bonds for 7, doing a multilink cube activity and the learning tasks in the textbook, followed by some reinforcement for memorization for number bonds through 7, then the next day on number bonds for 8, and so on. Or, you can introduce all the number bonds at once, and then spend several days with games and other activities to help your student memorize them.

Practice

Tasks 3-8, pp. 21-23

For tasks 3 and 5, get your student to come up with more than just one story when possible. For example, in task 3, there are 6 yellow and 1 green balloon, 3 small and 4 large balloons, 2 without stars and 5 with stars. Draw number bonds for the stories.

Workbook

Exercises 2-5, pp. 16-19

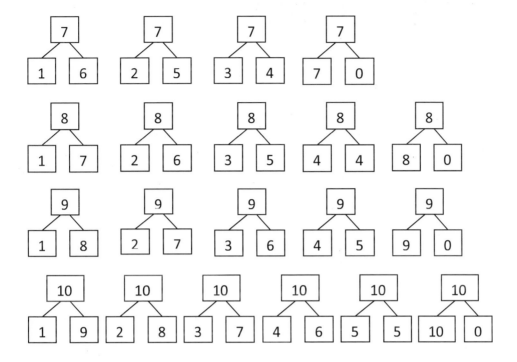

Reinforcement

Use 4 sets of number cards 1-10. Choose a target number between 6 and 10, which will be the whole. Shuffle the deck and place the whole deck face down. Start by turning over the cards, one at a time, and laying them out in a line in front of your student. As soon as she sees a pair anywhere that makes the target number, she can remove the pair from the line. Fill in the gaps with the next cards turned over. If you want no cards left over, only use cards up to and including the target number.

Game

Material: Playing cards 0-10

Goal: To make pairs of numbers that add up to the target number.

Procedure: Chose a target number, which will be the whole. Use only the cards up and including that number; for example, if the target number is 8, use cards 0-8. Shuffle and put 4 cards out on the table face up and the rest in a pile face down. Players take turns turning over a card from the face-down pile. If the card makes the target number with one of the cards on the table, that player keeps the two cards. Otherwise he leaves it face up on the table.

Enrichment

Use multilink cubes to find number bonds with more than two parts. Ask your student to split the cubes into more than 3 parts, and then write a number bond with more parts. See the first example below.

Write a number, ask your student to tell you two parts that make that whole. Then, for each part greater than 1, continue the process until there are only parts of 1. Ask your student to count the number of 1's at the end; the total 1's should be the same as the number you started with. See the second example below.

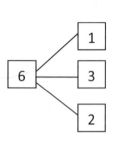

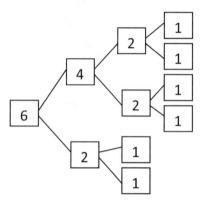

(3) Find the missing part of a number bond

Activity

Set out six to ten counters or other objects. Ask your student how many objects there are.

Cover up some of them and ask him how many are covered up. He can count on his fingers, if needed, by counting up from the number that is not covered up. Draw the number bond with a blank for the missing part, and fill in the missing part when he has the answer.

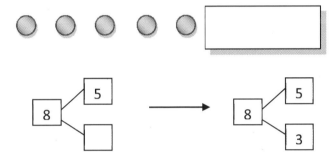

Repeat with other amounts.

Draw two circles. In one circle, put 5 objects. Write a number bond with 9 as the whole and 5 as one part.

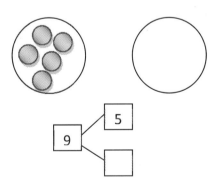

Ask your student to put more objects in the second circle to make 9. Fill in the missing part on the number bond.

Textbook

Task 9, p. 24

9. (a) 2
   (b) 6   (c) 0
   (d) 2   (e) 3
   (f) 6   (g) 9

Workbook

Exercise 6, pp. 20-22

1. Check answers.

2. (a) 2
   (b) 2
   (c) 6
   (d) 6
3. (a) 3 more
   (b) 2 more
   (c) 3 more
   (d) 4 more

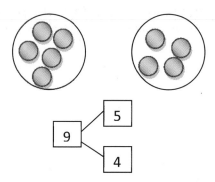

Repeat with other numbers.

Write a number bond with a missing part and ask your student to find the missing part. Allow him to use counters to find the answer, if needed.

Practice

Task 9, p. 24

Workbook

Exercise 6, pp. 20-22

Reinforcement

Use 4 sets of number cards, 0-10. Shuffle. Turn over two cards at a time. The larger number is the whole, and the smaller a part. Ask your student to name the other part.

(4) Practice number bonds that make 10

Activity

Use the dot cards described in the list of manipulatives at the beginning of this guide as Set 3, i.e., the dot cards for 0-10. Show your student the card for 10. You can leave it out as reference for a while. Mix up the remaining cards and show them to your student, one at a time. Ask him to tell you how many dots there are and how many more dots are needed to make 10.

Play the games shown in tasks 10 and 11, p. 25.

Task 10 is a game that can be played by two players. Each player has 10 counters. Players take turns dividing them into two sets, showing one set in an open hand and hiding one set in a fist. The other player must tell the number of objects that are hidden.

For Task 11 use number cards 0-10. For shorter games or activities use two sets, for longer practice use four sets. There are several ways you can use them.

1. Lay them all out face up, and ask your student to match tens.

2. Shuffle them. Show your student one at a time and ask her to tell you the number needed to make 10 with it.

3. Shuffle and put four cards out on the table face up and the rest in a pile face down. Players take turns turning over a card from the face-down pile. If the card makes 10 with one of the cards on the table, the player keeps the cards. Otherwise he leaves it face up on the table.

Discussion

Tasks 10-11, p. 25

Have your student supply the answers.

Workbook

Exercise 7, pp. 23-24

Textbook

Tasks 10-11, p. 25

10. 4

11. 0, 10; 1, 9; 2, 8;
    4, 6; 5, 5

Workbook

Exercise 7, pp. 23-24

1. (a) 3
   (b) 2

2. (a) 6   (b) 1
   (c) 5   (d) 4

3. 5 → 5
   4 → 6
   3 → 7
   2 → 8
   1 → 9
   0 → 10

4. (a) 9   (b) 7
   (c) 4   (d) 8

Reinforcement

Extra Practice, Unit 2, Exercise 1, pp. 9-12

Game

Material: Double-nine dominoes

Procedure: Turn the dominoes face down and mix them around. Turn one face up and put it in the center of the playing area. Players take turns taking a domino and matching one end to a domino on the table. The two matched ends must make a 10.

Test

Tests, Unit 2, 1A and 1B, pp. 7-10

Tests, Cumulative Test Units 1-2, A and B, pp. 11-16

Unit 3 – Addition

Chapter 1 – Making Addition Stories

Objectives

- Understand the meaning of **addition**.
- Make number stories for addition.
- Write addition equations using + and =.

Material

- Counters and other objects for addition stories: toys, pictures, etc.

Notes

In this chapter students will learn to write addition equations using the symbols + and = to represent the mathematical process of finding the total number of objects in two sets. The emphasis at this point is on understanding the meaning of addition, rather than on memorization of facts. Encourage your student to make up addition stories and represent the stories with an addition equation.

The '+' sign means to 'put together' and the '=' sign means "is the same as." Read addition equations in a variety of ways to your student, depending on the situation each represents. '5 + 2 = 7' can be read as, "five plus two equals seven" or "five and two is seven" or "5 and 2 more make 7" or "When we put 5 and 2 together, we get 7."

Though the curriculum at this level will write the two parts first, and the total, or whole, last, e.g. 2 + 5 = 7, there is nothing wrong with writing it the other way around: 7 = 2 + 5. The equal sign just means that both sides evaluate to the same number and does not dictate that what comes after it has to be the 'answer' to the addition problem. This is an important concept that most students are not exposed to until later. You might want to write some of the equations with the total first to get your student used to seeing it that way. In general, though, as we think through a problem, we write the calculations we must do first, 2 + 5, and then find the answer and write it.

There are three addition situations illustrated on p. 26.

1. Putting together: Put together 3 butterflies and 4 butterflies. There are 7 butterflies altogether.
2. Part-whole: The 6 red flowers and 4 yellow flowers are two parts of a whole. 10 is the whole. 6 and 4 make 10.
3. Adding on: Add 2 more children to the 3 children already in the sandbox. There will be 5 children altogether.

(1) Understand the meaning of addition

Textbook

Page 26

Workbook

Exercise 1, pp. 25-26

1. (a) 3
         4
         7
   (b) 4
        4
        8
   (c) 2
        4
        6

2. (a) 5
   (b) 9
   (c) 10
   (d) 7

Exercise 2, p. 27

1. (a) 2
        5
   (b) 1
        9
   (c) 3
        7
   (d) 4
        9

Activity

Illustrate the three addition situations using counters or other objects such as small toys . The names for the different situations are for your benefit; do not name them for your student. Make up stories to go along with the situations.

Part-whole:

Show 3 counters of one color and 6 of another. Ask your student for the number in each group, and then ask how many there are altogether. For example, there are 3 cardinals in the yard and 6 blue-jays. How many birds are there in the yard?

Putting together:

Show 5 objects in one group and 2 more of the same type in another group. Ask your student to count the number in each group. Then put them together into a single group. Ask him to tell you the total number. Say, "Putting 5 and 2 together make 7."

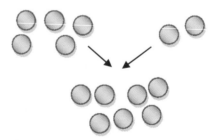

Adding on:

Draw a circle and put 4 counters in it, and 2 counters outside of it. Make up a story, such as; there are 4 frogs in the pond, and two on the bank. The two on the bank jump in the pond. Ask your student how many are in the pond. Move the counters from outside the circle into it, while she counts on from 4 to the total of 6. Say: 4 and 2 more make 6.

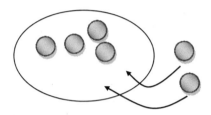

Discussion

Page 26

Ask your student to make up number stories to go along with this picture. For example, he can count the red flowers and the yellow flowers, and then tell you how many flowers there are altogether. Other possibilities include the butterflies on and off the flowers, the children in and coming to the sandbox, the number of boys and the number of girls.

Workbook

Exercise 1-2, pp. 25-27

Reinforcement

Use toys, pictures, or real-life situations to give your student more practice in creating addition stories.

(2) Relate addition stories to an addition equation

Textbook

Pages 27-28

Tasks 1-3, pp. 29-31

  3. Answers can vary.

Workbook

Exercise 3, pp. 28-30

1. (a) 8 + 1
   (b) 3 + 5
   (c) 6 + 1
   (d) 7 + 3

2. Answers can vary.
   (a) 5 + 3; 3 + 5; 6 + 2
   (b) 5 + 1; 4 + 2; 3 + 3

3. (a) 3 + 2 = 5
   (b) 3 + 3 = 6
   (c) 2 + 5 = 7
   (d) 5 + 4 = 9

Activity

Show your student 2 groups of the same objects, such as 7 counters in one group, and 2 in another group. Ask your student how many there are in each group.

Then move them together into one group and ask your student how many there are altogether. Tell her that we can show what is happening by writing an equation. Write: 7 + 2 = 9.

   7 + 2 = 9

Point to each part of the equation as you read it: 7 plus 2 equals 9.

Tell your student that '+' is called a plus sign. It means we are putting together the numbers on each side of it. The '=' means "is the same as" or "equals." The '7 + 2' on one side of the '=' sign is the same as '9' on the other side. Tell him we call a number sentence with an equal sign an **equation**. This type of number equation where we show that we are adding parts together is an **addition** equation.

Tell your student that 7 + 2 = 9 is a true statement, so it is a **fact**. When she sees an expression such as 7 + 2, and then gives the answer, she is giving an addition fact.

Show your student 2 sets of different objects, such as 4 counters of one color and 3 of another, or 4 multilink cubes and 3 counters. Ask him for the number in each group, and for the total. Say: 4 and 3 make 7.

Tell your student that since we are finding the total, we can also show what is happening by writing an addition equation. Write: 4 + 3 = 7.

$$4 + 3 = 7$$

Read the equation: 4 plus 3 equals 7.

Draw a circle and put 7 counters in it, and 3 counters outside of it. Ask your student how many are in the circle. Have your student count on as you add the 3 to the circle. Say: 7 and 3 **more** make 10. Tell her we also use the plus sign and write an addition equation to show that the first number is joined by more to make the total amount. Write: 7 + 3 = 10.

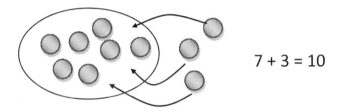

$$7 + 3 = 10$$

Repeat with other examples, asking your student to write the addition equation.

Set out some counters or toys that allow different stories for the same total. For example, set out 2 red and 1 green cube, and 2 red and 4 green counters, and see if your student can come up with several "stories" and write corresponding equations.

Discussion

Pages 27-28

Tasks 1-3, pp. 29-31

Workbook

Exercise 3, pp. 28-30

Reinforcement

Use toys, pictures, or real-life situations to give your student more practice in making number stories and writing addition equations. Try to find instances of each of the three types of addition situations.

Chapter 2 – Addition with Number Bonds

Objectives

- Relate addition stories to number bonds.
- Write two addition facts for a given number bond.
- Solve picture problems using addition.
- Memorize addition facts through 5, +0, +1.

Material

- Counters
- Other objects for addition stories: toys, pictures, etc.
- Fact game cards for addition facts within 5, +0, and +1

Notes

The number bonds learned in unit 2 will help your student master both the addition and subtraction facts within 10. These facts are associated with the part-whole concept of number bonds. Given a number bond we can write two related addition facts and two related subtraction facts.

$$5 + 3 = 8 \qquad 3 + 5 = 8$$
$$8 - 5 = 3 \qquad 8 - 3 = 5$$

Help your student commit the addition facts to memory. Some students can memorize these number facts easily with only a little drill; others may need additional help through games and structured activities. Some students, even those who excel at math comprehension, will always have trouble memorizing math facts, though they will probably be able to memorize the facts through 10 easily enough. Strategies for computing the facts through 20 from the facts through 10 will be given in a later unit.

At this stage most students already know addition facts within 5 and some easier addition facts such as +1. These are listed on the next page.

After this chapter help your student memorize these facts, and then add new facts as you go through succeeding lessons and units. Use flash cards, the suggested games in this guide, and computer games where you can set the facts that are to be practiced, etc.

Since first graders are still working on fine-motor skills, using worksheets, particularly timed ones, may not be a good idea, unless they are done orally. There are many web sites that will print out drill sheets to specification; you can use them and ask your student to look at the problem and answer orally if writing is an issue.

Try to use activities that do emphasize speed without stress so your student does move beyond doing all of them by counting on or using fingers.

There are mental math worksheets in the appendix you can copy and use. They can be used any time after they are listed in the lesson, and can be copied multiple times. You can time your student and then re-do the same sheet so she can see if her time improves.

0 + 0	0 + 1	0 + 2	0 + 3	0 + 4	0 + 5	0 + 6	0 + 7	0 + 8	0 + 9	0 + 10
1 + 0	1 + 1	1 + 2	1 + 3	1 + 4	1 + 5	1 + 6	1 + 7	1 + 8	1 + 9	
2 + 0	2 + 1	2 + 2	2 + 3							
3 + 0	3 + 1	3 + 2								
4 + 0	4 + 1									
5 + 0	5 + 1									
6 + 0	6 + 1									
7 + 0	7 + 1									
8 + 0	8 + 1									
9 + 0	9 + 1									
10 + 0										

You can take more than a day for any lesson. Take your time, and spend adequate time having your student review number bonds and practice the addition facts above.

(1) Add within ten by recalling the related number bond

Textbook

Page 32

    All answers are 8.

Tasks 1-2, p. 33

  1. 8; 8

  2. 7; 7

Workbook

Exercise 4, pp. 31-33

  1. (a) 8; 8, 6, 2
     (b) 6; 6, 1, 5
     (c) 10; 10, 3, 7
     (d) 5; 5, 2, 3

  2. 9   7
    9   7
    6   8
    6   8
    9   10
    9   10

  3. 6: 4 + 2; 3 + 3; 2 + 4
    8: 3 + 5; 4 + 4; 5 + 3
    5: 5 + 0; 1 + 4; 3 + 2
    7: 0 + 7; 4 + 3; 6 + 1
    9: 6 + 3; 5 + 4; 2 + 7

Exercise 5, pp. 34-35

  1. 8; 8
  2. 9; 9
  3. 8; 8
  4. 8; 8

Activity

Show your student 2 sets of different objects, such as 4 counters of one color and 3 of another, or 4 multilink cubes and 3 counters, or two sets of similar toys. Ask her to write the addition equation.

$$3 + 4 = 7$$

Tell your student that we have two parts and are finding the whole. So we can draw a number bond for this situation. One part plus the other part equals the whole.

$$3 + 4 = 7$$

Write a number bond. Ask your student to tell you an addition number story to go along with the number bond. Provide objects for props if that appeals to your student. Write two addition equations for the number bond. Tell your student that it does not matter which part we write first, the whole is still the same.

$$6 + 4 = 10$$
$$4 + 6 = 10$$

Repeat with another number bond. This time ask your student to write the two addition equations.

Discussion

Page 32

Practice

Tasks 1-2, p. 33

Workbook

Exercises 4-5, pp. 31-35

(2) Add within ten using number bonds

Discussion

Task 4, p. 34

Ask your student to make up an addition story for the two situations.

Activity

Tell your student an addition story involving a sum within 10. Draw a number bond and help her fill it in with the information in the story, and find the total. If she has memorized the number bonds, she should be able to find the total without needing objects, but if she has not, allow her to work out the answer with counters.

Repeat with other stories. Try to use each of the addition situations: part-whole, putting together, and adding on.

Workbook

Exercise 6, pp. 36-37

Reinforcement

Mental Math 1

Game

Material: Flash cards for addition facts through 5, +0, and +1 with answers on separate cards.

Procedure: Mix up the cards and put them all face down. Turn over the first card and put it in the middle. Players take turns turning over cards. If a card matches one that is face up, i.e. it is the answer to a fact or a fact for an answer, the player gets to keep both cards. Otherwise, he puts the card he turned over face up on the table.

Textbook

Task 3, p. 34

3. (a) 6
   (b) 10

Workbook

Exercise 6, pp. 36-37

1. (a) 7   (b) 7
   (c) 4   (d) 7

2. 9; 9

3. 10; 10

Chapter 3 – Other Methods of Addition

Objectives

- Count on to add 1, 2, or 3 to a number within 10.
- Learn and memorize addition facts through 10.

Material

- Counters
- Fact cards for addition through 10, with answers on separate cards (for matching games)
- Number cubes labeled with '+1', '+2', and '+3' on two sides each
- Number cards 0-10, 4 sets
- Dot cards for making 10, doubles, doubles + 1

Notes

This section introduces students to the "count on" strategy for addition. This strategy will help students work out harder addition facts before they can memorize them. Your student can use this strategy for +2 and +3 addition facts, and, if she knows doubles, with doubles + 1 facts (e.g., 3 + 4, 4 + 5).

Your student will probably be able to count on 1, 2, or 3 without using fingers. Fingers can be used if needed to begin with. Note that counting on as a strategy is used only for adding 1, 2, or 3, and should not be used when adding larger numbers. The goal is quick computation, and with adding on larger numbers it becomes harder to keep track of how many are added on and to know where to stop without fingers or number lines. Also, adding numbers where the sum is larger than 10 is taught in the context of the base-10 concept of our number system in Unit 6, rather than simply adding more numbers on. It is important that students think of numbers in the context of tens and ones, not just one more.

When your student does count on, be sure she is not including the number that she is counting on from. For example, for 6 + 3 she might say "6, 7, 8" and stop there, since she has counted up 3 numbers. Teach her to 'think' 6, and then count on 3 from there: 7, 8, 9.

Extra time is spent in this section on pairs of numbers that "make 10." If your student has learned the number bonds to 10, he might know the addition facts for 10 already.

The only addition facts within 10 not included in sums to 5, +1, +2, +3, and "make 10" are 4 + 4, 4 + 5, and 5 + 4. You can teach your student doubles (1 + 1, 2 + 2, 3 + 3, 4 + 4, 5 + 5). Then, 4 + 5, 5 + 4, as well as 3 + 4 and 4 + 3, are doubles + 1.

Dot cards (see the section on manipulatives) can help students initially with visualizing the addition facts for adding on to 5, the facts that make 10, doubles (e.g. 4 + 4), and doubles + 1 (e.g. 4 + 5 is the same as double 4 + one more). Students should be able to recognize numbers from dot patterns without having to count.

Continue with helping your student commit the addition facts to memory using games, drills, etc. After finishing this section, add the rest of the addition facts through 10 to those listed on p. 24 of this guide:

0 + 0	0 + 1	0 + 2	0 + 3	0 + 4	0 + 5	0 + 6	0 + 7	0 + 8	0 + 9	0 + 10
1 + 0	1 + 1	1 + 2	1 + 3	1 + 4	1 + 5	1 + 6	1 + 7	1 + 8	1 + 9	
2 + 0	2 + 1	2 + 2	2 + 3	**2 + 4**	**2 + 5**	**2 + 6**	**2 + 7**	**2 + 8**		
3 + 0	3 + 1	3 + 2	**3 + 3**	**3 + 4**	**3 + 5**	**3 + 6**	**3 + 7**			
4 + 0	4 + 1	**4 + 2**	**4 + 3**	**4 + 4**	**4 + 5**	**4 + 6**				
5 + 0	5 + 1	**5 + 2**	**5 + 3**	**5 + 4**	**5 + 5**					
6 + 0	6 + 1	**6 + 2**	**6 + 3**	**6 + 4**						
7 + 0	7 + 1	**7 + 2**	**7 + 3**							
8 + 0	8 + 1	**8 + 2**								
9 + 0	9 + 1									
10 + 0										

You can take more than a day for any lesson. Take your time, and spend adequate time having your student practice addition facts. Use your discretion for when to proceed. All the addition facts do not have to be mastered yet, and practice can continue through the next two units. Your student should be comfortable with all the addition facts through 10 by Chapter 2 of Unit 6.

(1) Count on to add 1, 2, or 3 to a number within 10

Textbook

Page 35

4 + 1 = 5
4 + 2 = 6
4 + 3 = 7

Tasks 1-4, pp. 36-37

1. 6 + 1 = 7

2. 7 + 2 = 9

3. 4 + 3 = 7

4. (a) 4   (b) 6
        5   8
        8   10
   (c) 5   (d) 8
        8   9
       10   10

Workbook

Exercise 7, pp. 38-39

1. (a) 5   (b) 9
   (c) 6   (d) 10

2. 7
   4
   5

3. 10
   7
   9

4. 6
   9
   10

Activity

Draw a number line for 0-10, or lay out number cards 1 to 10 in order. Write the expression: 5 + 1. Tell your student to put his finger on the number 5. Then tell him to hop forward 1 space, counting up as he does so: 6. Write the answer to the expression. Now write the expression: 5 + 2. Ask him to find the answer using the number line. Repeat with 5 + 3. Repeat with other addition problems within 10. If necessary, verify the answers by setting out corresponding numbers of counters under the numbers on the number line.

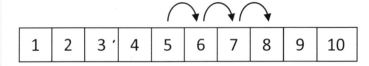

| 1 | 2 | 3 ʻ | 4 | 5 | 6 | 7 | 8 | 9 | 10 |

$$5 + 3 = 8$$

Write the expression: 6 + 3. Write the numeral 6 and then write 3 dots, or use 3 counters. Ask your student to supply the answer by using the dots or counters to count on from 6. Then write the expression: 3 + 6. Ask her to find the answer. Lead her to see that she can find the answer in the same way as with 6 + 3, not by counting on from 3 but rather by counting on from the larger number, 6. Repeat with other addition problems within 10, adding 1, 2, or 3.

6 + 3   6   6 + 3 = 9
                                                  3 + 6 = 9

Discussion

Page 33

Your student should start with 4, and then count on to add the frogs on the second lily pad.

Practice

Tasks 1-4, pp. 36-37

Workbook

Exercise 7, pp. 38-39

Reinforcement

Mental Math 2-3

Game

Material: Number line 0-10, with squares large enough for counters, counters, number cube labeled +1 on two sides, +2 on two sides, and +3 on two sides.

Procedure: Each player puts his or her counter at the beginning of the number line. Players take turns throwing the number cube and moving their counter on the number line by the amount face-up on the cube. If one player lands on another player's position, the other player goes back to 0. If the player is near 10 and the amount on the number cube causes the player to go past 10, the player goes back to 0. The first player to land on 10 wins.

(2) Learn addition facts for 10, doubles + 1

Textbook

Tasks 5-6, pp. 38-40

5. (a) 9   (b) 8
  (c) 7   (d) 6
  (e) 5   (f) 4
  (g) 3   (h) 2
  (i) 1   (j) 0

6. 10: 8 + 2, 4 + 6,
      3 + 7, 7 + 3,
      2 + 8, 9 + 1,
      5 + 5, 6 + 4
  9:   4 + 5, 7 + 2,
      6 + 3, 3 + 6,
      2 + 7, 0 + 9,
      5 + 4, 8 + 1

8. (a) 9 (b) 8 (c) 7
  (d) 6 (e) 6 (f) 10

Workbook

Exercise 8, p. 40

1. turtle

Exercise 9, p. 41

1. (a) 4   (b) 3
  (c) 6   (d) 9
  (e) 4   (f) 9
  (g) 5   (h) 2
  (i) 10   (j) 1
  (k) 7   (l) 5
  (m) 5   (n) 8
  (o) 10   (p) 9

Activity

Use counters or other small objects. Set out 6 of them in two rows. Write the expression: 3 + 3. Ask your student to find the answer. Then add another disk to one row. Write the expression: 3 + 4. Lead your student to see that he can find the answer from the answer to 3 + 3 by counting on 1.

       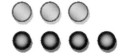

$$3 + 3 = 6 \qquad\qquad 3 + 4 = 7$$

Ask for the answer to 2 + 2, and then 2 + 3, by saying 2 + 3 is one more than 2 + 2.

Ask for the answer to 4 + 4, and then 4 + 5, by saying 4 + 5 is one more than 4 + 4. Illustrate with discs if necessary.

Give your student the dot cards for making 10 and ask her to write two addition equations for each. For example:

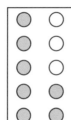

$$7 + 3 = 10$$
$$3 + 7 = 10$$

Practice

Tasks 5-8, pp. 38-39

Task 5 focuses on making 10, a major theme in this curriculum, and task 6 focuses on the addition facts for 8 and 9. Task 7 is a reminder to practice learning math facts. You may want to do these tasks and the workbook exercises after the following reinforcement activities.

Workbook

Exercise 8-9, pp. 40-4

Reinforcement

Mental Math 4-5

Extra Practice, Unit 3, Exercise 3A-3C, pp. 19-26

Games

Material: Number 0-10, 4 sets (or playing cards).

Procedure: Shuffle the cards and place them face down. Players take turns turning over two cards. If the two cards make a ten, they keep the cards; otherwise the cards go into a discard pile. When the cards are used up, the discard pile is shuffled and turned face down and the game resumes until all cards are used. The winner has the most cards.

Material: Flash cards for addition facts through 10 and answers on separate cards.

Procedure: Mix up the cards and put them all face down. Turn over the first card and put it in the middle. Players take turns turning over cards. If a card matches one that is face up, i.e. it is the answer to a fact or a fact for an answer, the player gets to keep both cards. Otherwise, he puts the card he turned over face up on the table.

Tests

Tests, Unit 3, 3A and 3B, pp. 25-28

Tests, Cumulative Test Units 1-3, A and B, pp. 29-36

Unit 4 – Subtraction

Chapter 1 – Making Subtraction Stories

Objectives

- Understand the meaning of **subtraction**.
- Make number stories for subtraction.
- Write subtraction equations using the symbols – and =.
- Write several subtraction equations for a given situation.

Material

- Counters
- Pennies
- Other objects for subtraction stories: toys, pictures, etc.

Notes

Subtraction is the reverse process of addition. Instead of adding on, we take away. When two parts or sets are put together, we add to find how many there are altogether. If one part is missing, we subtract to find out how many are in the missing part.

The '–' (minus) sign means to "take away." Read subtraction equations in a variety of ways to your student, depending on the situation they represent. For example, '8 – 3 = 5' can be read as, "eight take away three is five" or "eight minus three equals five" or "When we subtract three from eight, we get five."

In this chapter the emphasis is on understanding the meaning of subtraction, rather than the memorization of facts. Encourage your student to make up stories to illustrate subtraction.

There are two subtraction situations on p. 41.

1. Taking away: Take 3 carrots from 9; there are 6 carrots left. 8 birds are on a branch, three fly away (take away 3), there are 5 left.
2. Part-whole: There are 7 children. 2 are girls. How many are boys? Subtract to find out.

Another subtraction situation – finding the difference between two sets – will be taught in the first unit of *Primary Mathematics* 1B.

Illustrate subtraction concretely. In the teaching activities below, counters are used, but you can use other objects.

(1) Understand the meaning of subtraction

Activity

Illustrate some subtraction situations. Use counters or other objects.

Find the part left:

Show 7 objects in a group together. Make up a story about them, e.g., there are 7 cars.

Cover them up with an index card, or put something in front of them so your student does not see them. Move two objects out so he can see them. Tell your student that 2 went away, e.g., 2 cars drove off. Ask how many are left. After he answers, show how many are left.

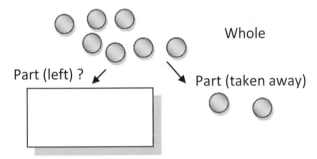

Whole

Part (left) ?

Part (taken away)

Find the missing part:

Show 5 objects in a group together. Use objects that are different on one side than the other, such as pennies or counters with a mark on one side. Then hide them from view and turn 2 of them over. Show the 2 you have turned over. Tell your student that you have 5 pennies, 2 of them have the tails facing up. Ask her how many have tails facing down (and heads up). After she answers, show the hidden ones.

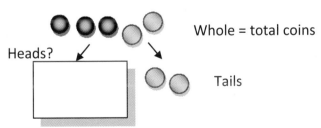

Whole = total coins

Heads?

Tails

Find the part taken away:

Show 6 objects in a group together. Make up a story about them, e.g., there are 6 dogs. Again hide them from your

Textbook

Page 41

Workbook

Exercise 1, pp. 42-43

1. (a) 2
      3
   (b) 2
      4
   (c) 1
      6
   (d) 4
      5

2. (a) 3
   (b) 3
   (c) 6
   (d) 7

Exercise 2, pp. 44-45

1. (a) 2
   (b) 4
   (c) 3
   (d) 3

2. (a) 4
   (b) 5
   (c) 4
   (d) 2

student in some way, and take some away. This time keep the part taken away hidden and show the part left. Tell your student that some went away (e.g., some dogs went away and 4 did not) and ask how many went away. After your student answers, show how many went away.

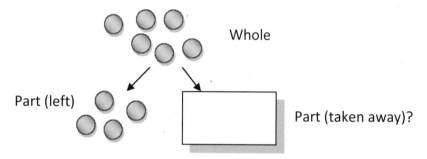

Discussion

Page 41

Ask your student to make up stories involving finding a part. For example, there were 8 birds on the branch. Three flew away. How many are left? Or, some flew away, and 5 stayed. How many flew away? There were 9 carrots. The bunnies ate 3 of them. How many are left? There are 7 children. 2 are girls. How many boys are there?

Workbook

Exercises 1- 2, pp. 42-45

(2) Relate subtraction stories to subtraction equations

Activity

Make up a subtraction story, e.g., "There are 9 balloons. Three of them popped. How many are left?" Illustrate with counters or other objects. Tell your student we can show what is happening by writing an equation. Write: $9 - 3 = 6$.

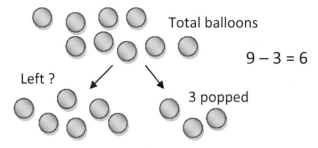

Point to each part of the equation as you read it: 9 take away 3 is 6.

Tell your student that '–' is called a minus sign and means we are taking away one part. The equal sign means that both sides mean the same thing, or are equal to each other. An equation with a minus sign and an equal sign is called a **subtraction equation**.

Repeat with other examples, similar to the examples in the previous lesson, using toys, pictures, or real-life situations. Ask your student to write the equations.

Discussion

Pages 42-43

Tasks 1-3, pp. 44-45

Workbook

Exercise 3, pp. 48-48

Reinforcement

Extra Practice, Unit 4, Exercises 1A-1B, pp. 29-34

Test

Tests, Unit 4, 1A and 1B, pp. 37-40

Textbook

Pages 42-43

Tasks 1-3, pp. 44-46

Answers will vary.

Workbook

Exercise 3, pp. 46-48

1. (a) $6 - 2$
   (b) $6 - 2$
   (c) $4 - 1$
   (d) $7 - 2$

2. (a) $6 - 4$
      $6 - 1$
   (b) $7 - 6$
      $7 - 3$

3. (a) $6-2=4$; $6-4=2$
   (b) $7-3=4$; $7-4=3$
   (c) $5-2=3$; $5-3=2$
   (d) $8-3=5$; $8-5=3$

Chapter 2 – Methods of Subtraction

Objectives

- Relate subtraction facts within 10 to the missing part of a number bond.
- Write two addition equations and two subtraction equations for a given number bond.
- Learn and memorize subtraction facts within 10.
- Review addition and subtraction within 10.
- Interpret addition and subtraction stories.

Material

- Counters
- Other objects for addition stories: toys, pictures, etc.
- Fact game cards for subtraction facts within 10
- Number cards 0-10, cards with '+', '–' , '='
- Number cube labeled '–1' on two sides, '–2' on two sides, and '–3' on two sides

Notes

As with addition facts, subtraction facts are associated with the part-whole concept of number bonds. Given a number bond we can write two related addition equations and two related subtraction equations.

$$5 + 3 = 8 \qquad 3 + 5 = 8$$
$$8 - 5 = 3 \qquad 8 - 3 = 5$$

As the student learned in the previous lesson, we write an addition equation to show that, given the parts, we are finding the whole:

$$5 + 3 = 8 \qquad 3 + 5 = 8$$

If we are given the whole, or total amount, and one part, we use a subtraction equation to show that we are finding a missing part.

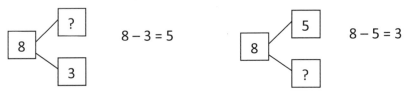

$$8 - 3 = 5 \qquad\qquad 8 - 5 = 3$$

In this chapter your student will write a family of four related addition and subtraction facts. She needs to realize that we use an addition equation to put two parts together, so the answer will be larger than each part. We use a subtraction equation to show that we start with a whole, or total, and know one part, but need to find a missing part. The missing part can either be the part taken away, or the part left. The answer to a subtraction equation is always smaller than the whole, and we subtract from the larger number. This concept is reinforced in the exercises in which the student has to insert the missing operand ('+' or '−').

Counting back can help students work out the subtraction facts for subtracting 1, 2, and 3 before they memorize them. Counting back 1, 2, 3 is doable without manipulatives, but counting back larger numbers than 3 usually needs some way to keep track of how many numbers have been counted back, such as fingers or a number line. Since counting back is meant to help students calculate mentally, the "count back" strategy is restricted here to counting back 1, 2, or 3.

When counting back, make sure your student does not make the common error of including the number that is being counted back from. He needs to begin with that number, but count back from there. Introducing the strategy with a number line can help your student not make this error, since he counts the number of hops.

Another strategy students can use to find the answer to a subtraction equation is to recognize when two numbers are 1, 2, or 3 apart. If the numbers are not too far apart, it is easy to count up from the part to the whole to find the answer. For example, in $8 - 6$, start at 6, count up 2 to get to 8, so $8 - 6 = 2$. Task 9 in the textbook includes numbers that are 1 or 2 apart, but this strategy can be used for numbers that are 3 apart as well, as most students can keep track of counting up 3 without manipulatives.

Students should be thoroughly familiar with pairs of numbers that make 10 by now, and so be able to answer problems where a number is subtracted from 10. For students who make pictures in their minds, visualizing the 5 by 2 dot pattern on a dot card can be useful in helping them with subtraction from 10.

Help your student commit the subtraction facts to memory. After lesson 1 begin with facts within 5, which your student should be able to master quickly by recalling number bonds. After lesson 3 you can include subtracting 1, 2, or 3 from numbers within 10. After lesson 4 include numbers that differ by 1, 2, or 3, and the rest of the subtraction facts for 10. You can work with these facts a little each day as you continue with the lessons in the textbook. If your student has trouble you can divide the facts up into the following groups and work with specific groups before combining.

Subtract equal numbers:

$0-0$ $1-1$ $2-2$ $3-3$ $4-4$ $5-5$ $6-6$ $7-7$ $8-8$ $9-9$ $10-10$

Subtract 0:

$0-0$ $1-0$ $2-0$ $3-0$ $4-0$ $5-0$ $6-0$ $7-0$ $8-0$ $9-0$ $10-0$

Subtract 1:

$1-1$ $2-1$ $3-1$ $4-1$ $5-1$ $6-1$ $7-1$ $8-1$ $9-1$ $10-1$

Subtract 2:

$2-2$ $3-2$ $4-2$ $5-2$ $6-2$ $7-2$ $8-2$ $9-2$ $10-2$

Subtract 3:

$3-3$ $4-3$ $5-3$ $6-3$ $7-3$ $8-3$ $9-3$ $10-3$

Subtract a number that is one less:

$1-0$ $2-1$ $3-2$ $4-3$ $5-4$ $6-5$ $7-6$ $8-7$ $9-8$ $10-9$

Subtract a number that is two less:

$2-0$ $3-1$ $4-2$ $5-3$ $6-4$ $7-5$ $8-6$ $9-7$ $10-8$

Subtract a number that is three less:

$4-1$ $5-2$ $6-3$ $7-4$ $8-5$ $9-6$ $10-7$

Remaining facts:

$8-4$ $9-4$ $9-5$ $10-4$ $10-5$ $10-6$

    Use computer games where specific facts can be set, or play variation of the suggested games in the previous section. Use activities or games where the student sees the fact in the form of an expression with the minus sign, rather than the type of games suggested in the number bond section, unless your student still needs practice with number bonds. By now your student should be learning to look at the abstract expression and find the answer without pictures or manipulatives, even if he is thinking pictures in his head to help find the answer. Eventually mix in both addition and subtraction facts.

    In lesson 6 your student will learn how to determine whether the answer to a math story requires addition or subtraction. This can be a challenge for many students. At this stage, most of the story is in the form of a picture. Do not try to get your student to rely on key words. Study the problems on pages 61-66 in the workbook On pages 61 and 62 the word altogether is found in the stories that require addition, but the second problem on page 64 uses the word altogether in a subtraction story. The stories do not deliberately use words like "left" or "taken away" to cue the student to a subtraction equation.

In helping your student determine whether to add or subtract, ask leading questions to get her to think in terms of part and whole, or total. Does the story give two parts and ask for a total amount? Does the story give a total amount, and a part, and ask for another part?

The pictures will help the student in the workbook exercise, but in later exercises, the student will not have a picture. Encourage your student to act out the story, or use counters to represent the story, or draw a picture, if she is having difficulty deciding whether to add or subtract.

In some stories, your student might want to write an equation in the form $3 + ? = 7$, where ? is the number that is the answer to the question posed by the story, rather than a subtraction equation. During lesson time ask your student to write a subtraction equation. Your student needs to be able to write subtraction equations, and subtraction emphasizes that we are starting with a whole. Writing the subtraction expression is simply a way of recording that, even if it is solved mentally by counting up from the smaller number. Later students will learn how to turn a missing addend equation into a subtraction equation. For example, an equation such as $312 + \underline{\quad} = 432$ is solved with subtraction: $\underline{\quad} = 432 - 312$, and in algebra, when showing the steps for the solution, subtraction is used: $312 + n = 432$; $n = 432 - 312$.

(1) Subtract within ten by recalling the related number bond

Textbook

p. 47

6 – 2 = 4	6 – 1 = 5
6 – 4 = 2	6 – 5 = 1
6 – 6 = 0	6 – 3 = 3
6 – 0 = 6	

Tasks 1-3, pp. 48-49

1. 4; 4

2. 3; 3

3. (a) 6
   (b) 3

Workbook

Exercise 4, pp. 49-50

1. 3; 2; 3
   3; 4; 1; 3
   0; 6; 0; 6
   2; 9; 2; 7

2. 4   3
   6   3
   7   6

Exercise 5; pp. 51-52

1. 5; 5

2. 2; 2

3. 5; 5

4. 6; 6

Activity

Set out 8 counters or other objects. Make up a story, such as: "There are 8 cookies. We set 5 aside to eat later." Move 5 away from the 8. Ask your student how many are left. Draw a number bond to represent this, writing 8 and saying that this is the total number of cookies, then 5 and saying that this is the number we set aside. Then write a question mark in the other part of the number bond. Tell your student this is what we want to find, the missing part of the number bond. We use the minus sign to show that we want to find a missing part. Write both an addition and subtraction equations with a question mark for the missing part. Then write the subtraction equation with the answer.

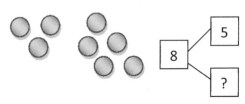

$$8 + ? = 5$$
$$8 - 5 = ?$$
$$8 - 5 = 3$$

Draw some number bonds with a missing part and ask your student to write the subtraction equation and the answer.

$$7 - 4 = 3$$

Discussion

Page 47

Tasks 1-3, pp. 48-49

You can ask your student to make up stories for Task 3.

Workbook

Exercises 4-5, pp. 49-52

(2) Write a family of addition and subtraction equations

Activity

Show your student two sets of items, such as toy cars and airplanes, cubes and counters, or two colors of counters. Make up a story for an addition equation, such as, "Toby has 3 cars and 4 planes. How many toys does he have altogether?" Ask your student to write two addition equations. Then make up a story such as "Toby has 7 cars and planes. 3 of them are cars. How many are planes?" Ask your student to write the subtraction equation. Repeat with the related situation, e.g., "Toby has 7 toy cars and planes. He put all his 4 planes away. How many cars does he have out still?" Point out that there are now four equations, two for addition and two for subtraction.

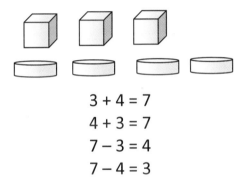

$3 + 4 = 7$

$4 + 3 = 7$

$7 - 3 = 4$

$7 - 4 = 3$

Write an addition or subtraction equation. Ask your student to draw a number bond for it, and then write the other three related addition or subtraction equations.

$5 - 4 = 1$

$1 + 4 = 5$

$4 + 1 = 5$

$5 - 1 = 4$

$5 - 4 = 1$

Discussion

Task 4, p. 50

Workbook

Exercises 6-7, pp. 53-56

Textbook

Task 4, p. 50

4. (a) $3 + 2 = 5$ $5 - 2 = 3$
        $2 + 3 = 5$ $5 - 3 = 2$
   (b) $7 + 2 = 9$ $9 - 2 = 7$
        $2 + 7 = 9$ $9 - 7 = 2$

Workbook

Exercise 6, pp. 53-54

1. (a) 6   6
         1   5
   (b) 7   7
         3   4
   (c) 5   5
         3   2

2. (a) $5 + 2 = 7$ $2 + 5 = 7$
        $7 - 2 = 5$ $7 - 5 = 2$
   (b) $6 + 4 = 10$ $4 + 6 = 10$
        $10 - 6 = 4$ $10 - 4 = 6$
   (c) $6 + 3 = 9$ $3 + 6 = 9$
        $9 - 3 = 6$ $9 - 6 = 3$

Exercise 7, pp. 55-56

1. –   –
   +   –
   –   +

2. Answers may vary.
   $10 - 3 = 7$
   $3 + 4 = 7$
   $6 + 0 = 6$
   $9 - 2 = 7$
   $8 - 5 = 3$

Reinforcement

Set out three number cards that belong in a number bond, and two sign cards, an = sign and either + or −. Ask your student to arrange them in an equation. Repeat with other cards.

Set out six number cards for two different number bonds, two '=' cards, a '+' card, and a '−' card and ask your student to use the cards to make up two number equations. Repeat, this time with the cards needed to make three number equations.

Mental Math 6

(3) Count back to subtract 1, 2, or 3 from a number within 10

Activity

Draw a number line for 0 to10, or lay out number cards 1 to 10 in order. Write the expression 8 – 1. Tell your student to put her finger on the number 8. Then tell her to hop backwards 1 space, counting down as she does so: 7. The answer to 8 – 1 is 7. Write the answer to the expression. Now write the expression 8 – 2 and ask her to hop backwards two spaces from 8 and tell you the answer. Then repeat with 8 – 3. Repeat with other subtraction problems within 10.

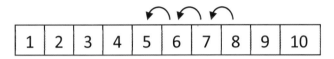

| 1 | 2 | 3 | 4 | 5 | 6 | 7 | 8 | 9 | 10 |

$$8 - 3 = 5$$

Write an expression such as: 7 – 3. Draw a circle and put 7 counters in it. Ask your student to count them, and then three counters aside one at a time, counting back as you do so. Point out that the number you say last is the answer to 7 – 3. Have your student do the same with other subtraction problems within 10, subtracting 1, 2, or 3. Emphasize that when we count back to subtract we don't include the number we start with.

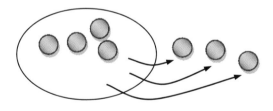

$$7 - 3 = 4$$

Discussion

Tasks 5-8, pp. 51-52

Workbook

Exercise 8, pp. 57-58

Textbook

Tasks 5-8, pp. 51-52

5. 4

6. 5

7. 7

8. (a) 3   (b) 3
       5   6
       8   8
   (c) 3   (d) 2
       4   5
       8   6

Workbook

Exercise 8, pp. 57-58

1. (a) 4   (b) 5
   (c) 6   (d) 7

2. 5; 3; 7

3. 5; 3; 7

4. 2; 5; 7

Reinforcement

Mental Math 7

Game

Material: Number line 0-10, with squares large enough for counters, counters, number cube labeled '–1' on two sides, '–2' on two sides, and '–3' on two sides.

Procedure: Each player puts his or her counter at the end of the number line on 10. Players take turns throwing the number cube and moving their counter. If one player lands on another player's position, the other player goes back to 10. If the player is near 0 and the amount on the number cube causes the player to go past 0, the player goes back to 10. The first player to land on 0 wins.

(4) Recognize numbers that differ by 1 or 2, subtract from 10

Activity

Draw a number strip of 10 squares labeled with the numbers 1 to 10, or lay out number cards 1 to 10 in order. Write the expression: 8 – 6. Ask your student to locate the two numbers on the number line. Point out that these numbers are close to each other. Ask her to find out how close by counting on from 6 to 8. Write: 8 – 6 = 2. Tell your student than when two numbers are close to each other in a subtraction equation; it is easy to find the answer by counting on from the smaller number to the larger number. You can put two colors of counters below the line to show that by counting up from the smaller number to the whole we are finding the missing part.

1	2	3	4	5	6	7	8	9	10

$$8 - 6 = 2$$

Repeat with numbers that are 1, 2, or 3 apart, writing the subtraction expression and having your student count up to find the answer.

Give your student the dot cards for making 10 and ask him to write two subtraction equations for each. For example:

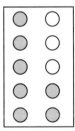

$$10 - 7 = 3$$
$$10 - 3 = 7$$

Discussion

Tasks 9-11, p. 53

If your student remembers all the ways to make 10, tasks 10 and 11 will be easy. With number bonds, you can remind your student that we are simply finding the missing part, as on p. 38 of the text.

Textbook

Task 9-11, p. 53

9. (a) 0   (b) 0
     1   1
     2   2
   (c) 1   (d) 2
     1   2
     1   2

10. 4

11. (a) 7   (b) 5
    (c) 2   (d) 1

Workbook

Exercise 9, p. 59

1. mountain, houses, field, sky and river

Reinforcement

Mental Math 8-9

Game

Material: Flash cards for subtraction facts through 10 and answers on separate cards.

Procedure: Mix up the cards and put them all face down. Turn over the first card and put it in the middle. Players take turns turning over cards. If a card matches one that is face up, i.e. it is the answer to a fact or a fact for an answer, the player gets to keep both cards. Otherwise, he puts the card he turned over face up on the table.

(5) Review addition and subtraction facts

Activity

Use drill sheets, fact cards, or games to practice the addition and subtraction facts through 10. You can use the mental math pages at the back of this guide. You can give your student the same sheet each day, and use a timer to see how far he gets in 3 minutes. If your student has trouble writing, do them orally. You can also see how many fact cards he can get correct in 3 minutes. Fact cards can be mixed up each time, or you can select only the facts that are causing difficulty.

Discussion

Tasks 12-14, pp. 54-55

Task 13 is a game suggestion where the student matches the fact cards with the answer. A 0 does need to be included, though one is not shown in the first printing of the textbook.

Workbook

Exercises 10 and 13, pp. 60 and 67

Reinforcement

Use playing cards 1-10, or 4 sets of number cards 1-10, and 5 cards each with '+',' −', and '='. Shuffle the cards, turn the deck face down, and ask your student to draw one card at a time and lay them out on the table face up. Every time she lays one down, she should look over the cards to see if she can spot a set of 3 numeral cards, a plus or minus card, and an equal card that will make an addition or subtraction equation. If so, she can remove them.

Mental Math 10

Game

Material: Playing cards 1-10, or 4 sets of number cards 1-10, and 5 cards each with '+',' −', and '='.

Procedure: Shuffle the cards and place them face down. Players take turns drawing cards one at a time and seeing if they can form an addition or subtraction equation with the cards drawn. If so, the player should set the cards down showing the equation. The goal is to end of with the fewest cards still in hand.

Textbook

Tasks 12-14, pp. 54-55

12. train 5: $9 - 4$;
    $8 - 3$; $2 + 3$; $5 - 0$
    train 6: $7 - 1$;
    $10 - 4$; $0 + 6$; $9 - 3$
    train 7: $8 - 1$;
    $3 + 4$; $10 - 3$; $9 - 2$

13. (a) 4   (b) 1   (c) 3
    (d) 6   (e) 0   (f) 6

Workbook

Exercise 10, p. 60

1. 5: cross out
   $8 - 1$; $3 + 1$
   6: cross out
   $4 + 0$; $5 + 2$
   7: cross out
   $10 - 7$; $2 + 4$
   8: cross out
   $8 - 1$; $6 + 3$

Exercise 13, p. 67

1. (a) 6   (b) 1
   (c) 3   (d) 1
   (e) 8   (f) 1
   (g) 5   (h) 7
   (i) 0   (j) 10
   (k) 5   (l) 4
   (m) 3   (n) 0
   (o) 1   (p) 6

(6) Interpret addition and subtraction stories

Activity

Tell your student a simple math story that requires addition or subtraction to solve. Use the problems in workbook exercises 11 and 12, pp. 61-66 as a guideline, or draw on real-life situations. For example:

➤ There were 8 dogs in our yard. 5 of them went into the neighbor's yard. How many are now in our yard?

➤ I saw 4 birds at our bird feeder this morning, and then I saw 2 squirrels there. How many animals did I see at the bird feeder this morning?

➤ There were 10 deer in the field. 7 were adults and the rest were fawns. How many fawns were there?

Ask your student to write an equation for the story. It is likely your student will be able to answer the problems easily, but writing the appropriate equation is more difficult. As numbers get larger, being able to write an equation becomes more important. If she has difficulty, provide her with counters or other objects so she can act out the story and determine whether she is finding the total or a missing part, and therefore whether to write an addition or a subtraction equation. Or, ask her to diagram the situation — circles could represent the dogs, for example.

After your student has found the answer, ask him to answer the question in the story in a complete sentence.

Workbook

Exercises 11-12, pp. 61-66

Some students will find the word problems easier to understand if you read them aloud for the student.

Reinforcement

Write an addition or subtraction equation and ask your student to make up a story to go with it.

Extra Practice, Unit 4, Exercise 2A-2C, pp. 35-42

Mental Math 11

Workbook

Exercise 11, pp. 61-63

1. +; 7; 7
2. −; 3; 3
3. +; 10; 10
4. −; 4; 4
5. −; 3; 3
6. +; 4; 4

Exercise 12, pp. 64-66

1. $4 + 3 = 7$; 7
2. $10 - 4 = 6$; 6
3. $7 - 2 = 5$; 5
4. $2 + 4 = 6$; 6
5. $4 + 4 = 8$; 8
6. $8 - 3 = 5$; 5

Challenge

Students often think that two numbers go on the left side of the '=' sign, with '+' or '−' between them, and the answer goes on the right side of the = sign. Encourage your student to understand that the '=' sign simply means that the expressions on both sides are equal by having her solve equations such as those shown below. Allow her to use counters or multilink cubes, if necessary.

$3 + 7 = 5 + \square$

$5 - 4 = 6 - \square$

$3 + 2 = 10 - \square$

$9 - \square = 3 + 3$

$4 \bigcirc 3 = 10 \bigcirc 3$

$8 \bigcirc 2 = 3 \bigcirc 3$

$5 \bigcirc 2 = 9 \bigcirc 2$

Tests

Tests, Unit 4, 2A and 2B, pp. 41-44

Tests, Cumulative Test Units 1-4, A and B, pp. 45-51

Unit 5 – Position

Chapter 1 – Position and Direction

Objectives

- Use position words.
- Use left and right in directions.

Material

- Various objects that can be moved around, e.g. toys, multilink cubes
- Square graph paper (you can copy the one in the appendix, or make a larger one)

Notes

In this chapter students will focus on words that describe position – under, above, below, up, down, far, near, next to, behind, in front of, up, down, left, right. These words will then be used later to describe geometric shapes. Students will also follow directions for movement on a simple grid.

Your student probably already knows these words but may not use them precisely. Use them with actual object in her surroundings before looking at pictures in the textbook or elsewhere. Since math books show pictures of 3-dimensional objects, your student needs to interpret 2-dimensional flat pictures as representing 3-dimensional objects and position.

Make sure your student knows left and right. In the next chapter, he will be locating objects in a line based on position from the left or from the right.

There are various ways to help your student remember which side is left and which right. Your student could associate left or right with which hand she writes with. Or, the left hand, when held up palm outward, with the thumb at right angles to the fingers, forms the letter L, the first letter in "left". This trick, though, requires the student to remember which way a capital L goes.

You may want to teach your student to distinguish between his or her right and someone else's right when the other person is facing the student. However, for any pictures on a grid in this curriculum, such as the one on p. 70 of the workbook, the student should assume that right and left are according to her orientation (not the rabbit's, which is facing out from the page).

(1) Use position words

Activity

Discuss the position of various objects in the room, using the words under, above, below, up, down, far, near, next to, behind, in front of, up, and down. Get your student to describe the position of a specific object in relation to objects around it; e. g., the ball is under the table.

Get your student to distinguish between her right hand and her left hand. Place some objects in a row and ask her questions such as: "What is to the left of the block?"

Give your student some objects and tell her to place them according to specific directions. For example: Put the block under the doll, the car to the right of the block, the bear behind the doll, and the truck far away.

Discussion

Pages 56-57

Tasks 1-3, pp. 58-59

Make sure your student understands what a 'step' is on the graph in task 3, and that 'up' is towards the top of the page. Get him to give alternate directions to the zoo in (d).

Workbook

Exercise 1, pp. 68-70

Reinforcement

Extra Practice, Unit 5, Exercise 1, pp. 47-48

Provide your student with square graph paper and let him draw a map, such as a treasure map, with landmarks at intersections. Ask him to give you directions to the treasure from different places on the map.

Show your student a street map and discuss the location of familiar places, and the directions to those places.

Tests

Tests, Unit 5, 1A and 1B, pp. 53-56

Textbook

Pages 56-57

Tasks 1-3, pp. 58-59

1. 2; the blue bird
2. left
3. (a) store
   (b) park
   (c) 3; 1
   (d) Answers can vary.

Workbook

Exercise 1, pp. 68-70

1. above
   behind
   below
   next to
   in front of
2. (a) starfish
   (b) Yes
   (c) fish
   (d) eel
   (e) crab
3. (a) No
   (b) Lettuce
   (c) Carrot
   (d) Right
   (e) Down

Chapter 2 – Ordinal Numbers

Objectives

- Name a position using ordinal numbers 1st through 10th.
- Find an ordinal position from the left or from the right.

Material

- Cards with the following: 1st, 2nd, 3rd, 4th, 5th, 6th, 7th, 8th, 9th, 10th (see appendix)
- Ten objects that can face in a particular direction, e.g. toy animals, cars, people

Notes

Ordinal numbers (first, second, third, fourth, etc.) are used to denote relative position. Your student should see the difference between ordinal numbers (1st, 2nd, 3rd, 4th, etc.) and counting, or cardinal, numbers (1, 2, 3, 4, etc.). For example, he should recognize the difference between counting out and coloring 3 items versus indicating the third or coloring the third item in a row. He also needs to know right from left, and that if we are not told which side to count from to find an item in a specified position, we start on the left (starting and going in the same direction we write or read).

Most children will pick up the concept of ordinal numbers used for position, particularly first and last, but usually also second and third. "I want to be first!" "I came second." "I got the last one." Ordinal numbers are also used to indicate sequence of events, though usually not past third, or even past first, with the word "next" being used instead.

Position and ordinal numbers can be used for logic types of problems. If you use one of the supplementary books, there will be problems such as the following:

➢ In a race, John is second and Mary is ninth. How many people are between John and Mary?

➢ Jasmine is fourth from the right and second from the left in a row. How many are in the row?

These can easily be solved using drawings or counters.

(1) Name position using ordinal numbers

Activity

Set out objects, such as toy animals, people, or cars, in a row. Ask your student to imagine they are in a race and talk about which one comes in first, second, third, etc. Then set them in a line with all of them facing forward to the front of the line, and, talk about which one is first in line, second in line, etc. Ask him to point out which one is fourth, seventh, etc. Show him cards with 1st, 2nd, 3rd, 4th, ..., 10th written on them, and ask him to put them in order. Pick up a card, such as 5th, and ask him to pick out the corresponding object in the line.

Now set the objects in a line from left to right in front of your student, with each facing the student so that it does not appear they are following each other. Point out the object farthest to the left and tell her that it is first, and ask her to find the one that is "5th from the left." Repeat with other positions. Point to one of the objects and ask her to tell you what position it is in, e.g., "The red car is 4th from the left." Repeat using positions from the right. Then, mix up the directions, e.g. "Which one is 3rd from the right? The bear is third from which side? Pick out the 2nd toy from the left. Pick out 2 toys."

Tell your student that when we are not told what side to start at, we say the one to the farthest left is first.

Discussion

Page 60

Tasks 1-2, p. 61

Workbook

Exercises 2-3, pp. 71-75

Reinforcement

Extra Practice, Unit 5, Exercise 2, pp. 49-50

Test

Tests, Unit 5, 2A and 2B, pp. 57-60

Textbook

Page 60

Tasks 1-2, p. 61

1. B
   J
   first
2. (a) giraffe
   (b) tiger

Workbook

Exercise 2, pp. 71-73

Check answers

Exercise 3, pp. 74-75

1.-5. Check answers

6. (a) 4th, 1st, 5th, 2nd, 3rd
   (b) 2nd, 5th, 4th, 1st, 3rd

Review

Note

Before continuing on to the next unit, make sure your student knows the addition and subtraction facts through 10 well, particularly pairs of numbers that make 10.

The workbook has three review exercises in succession. You can use them as assessments to see if you need to go back and re-teach any part of units 1-5, or save them and ask your student to do several problems a day as you go on as a form of continuous review.

Tests

Tests, Cumulative Test Units 1-5, A and B, pp. 61-68

Workbook

Review 1, pp. 76-79

1. 7   10
2. four
   two
   five
   ten
   three
3. 6   7
   9   0
4. eight   five
5. check answers
6. 7   2
7. (a) 3 and 5
   (b) 2 and 8
8. (a) 10   (b) 5
9. 9   6   0
10. draw 5 flowers
11. (a) draw 8 flowers
    (b) 10

Review 2, pp. 80-83

1. 9
   9
   7
   5
2. $3 + 2 = 5$; $2 + 3 = 5$;
   $5 - 3 = 2$; $5 - 2 = 3$
3. (a) 5   (b) 1
   (c) 10   (d) 0
   (e) 7   (f) 9
   (g) 8   (h) 4
   (i) 9   (j) 6   (continued next page)

Workbook

Review 2, pp. 80-83 (continued

4. 4 + 5 = 9

9

5. 7 − 2 = 5

5

6. 8 − 2 = 6

6

7. 8 − 2 = 6

6

8. 3 + 4 = 7

7

Review 3, pp. 84-88

1. (a) eight

(b) seven

(c) five

(d) nine

2. first set

3. second set

4. 8 → 2   4 → 6   0 → 10

5. 5th; 3rd; (1st;) 4th; 2nd

6. (a) 7   (b) 9

(c) 5   (d) 6

(e) 6   · (f) 5

7. (a) −   (b) +

(c) +   (d) −

(e) −   (f) +

(g) −   (h) −

8. 6 + 2 = 8

8

9. 9 − 4 = 5

5

10. post office

flag

school

library

Unit 6 – Numbers to 20

Chapter 1 – Counting and Comparing

Objectives

- Count to 20 by building up from 10.
- Read and write numerals and number words for 11 to 19.
- Relate numbers 11 to 20 to a ten plus ones.
- Count backwards from 20.
- Determine missing numbers in a sequence from 1 to 20.
- Compare and order numbers within 20.

Material

- Number cards 0-20
- Number cards for 10 and ones half that size for 1-9
- Number word cards, eleven to twenty
- Dot cards 1-20
- Counters
- Multilink cubes
- 4 sets of number cards, 0-10

Notes

In this unit students are introduced to the base-ten system for writing numbers. We use 10 **digits**, 0-9. Only 1-9 are whole numbers. With these digits, we can write the numerals to represent quantities of one through nine. To write ten, we write a 1 in the next "place" to the left: 10. This is called the tens place, and the 1 there tells us that we have one ten. The 0 in 10 is in the ones place, and tells us there are no ones left over after the counted items are placed in groups of ten. The next number, 11, is written by starting over again with 1 in the ones place. So 11 is 1 ten and 1 one. The number in the tens place tells us how many tens (groups of ten ones) we have, and the number in the ones place tells us how many ones (groups of 1) we have. In *Primary Mathematics* 2 students will learn that 10 tens can also be grouped, this time into hundreds.

Many first graders can count past ten and beyond 100 by simply counting on. Here, students will count objects up to 20 by first making a ten and then writing down the number of tens followed by the number of ones left.

In *Primary Mathematics 1B* students will learn numbers to 100. Again, the emphasis will be on place-value, and using place-value concepts for addition and subtraction, not simply counting on.

Some students get confused which number is written first for "teen" numbers since the ones are said first, e.g. "eighteen". You can ask your student to read the number as "ten and eight" and later work on relating the number words to the numbers.

In this chapter students also read and write number words. If your student's writing and spelling abilities are not up to writing number words at this time, don't require it. In places in the workbook where the student is asked to write the number in words, which requires spelling them, you can either ask him to just write the numerals, or provide the words to copy. Decide for yourself how much time you want to spend on requiring him to write the number words at this time and adjust the workbook exercises accordingly. Reading, writing, and spelling of number words can be taught apart from math lessons, and spelling number words can be taught later than reading them. Focus on reading the number words for now.

(1) Count to 20 by building up from 10

Textbook

Pages 62-63

Task 1, p. 64

1. (a) 16
    (b) 15
    (c) 18

Workbook

Exercise 1, pp. 89-91

1. 12   16   11
   17   14   15

2. 17   13
   14   20
   15   16
   11   12

3. 16   15
   14   12
   13   11

Exercise 2, pp. 92-93

1. 11; eleven
   16, sixteen
   13, thirteen
   17, seventeen
   20, twenty

2. 16   13   11
   12   20
   17   14
   19   18
   15

Activity

Write the numbers 0 through 9. Tell your student that these are called **digits**.

Set out 9 multilink cubes and ask your student to count them. Then add another cube and ask her how many there are. Ask her to link them together in a row. Tell her we now have a group of ten cubes in a 10-rod. Set out another cube to the right of the 10-rod, and ask your student how many there are. Write the number 11. Tell her that we write 11 as 1 group of ten, and 1 more. 10 and 1 is 11. You can show this by sliding a '1' number card over the '0' of a '10' card.

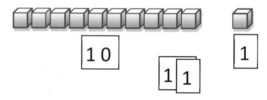

Take away the single cube and write 10 again. Lead your student to see that 10 means 1 group of 10, and 0 more. The first digit means there is 1 ten, and the second means there are no more.

Now add 2 more cubes next to the 10-rod, and ask your student to write the number. Guide him in writing a 1 for the group of ten, followed by 2 for how many more than the groups of ten. Ask him to read the number (if necessary, he can count on from ten: eleven, twelve. 12 is read as twelve.) Tell him that twelve is 10 and 2, pointing to the digits. Use number cards for 10 and 2 and slide the 2 over the 0 of the 10.

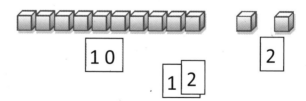

Continue with 13, 14, 15, 16, 17, 18, and 19. Give your student the card for 10 and the half-sized cards for the digits and have him form the numbers himself

After 19 add one more and ask your student how many there are. Point out that 20 is 10 and 10 more. Tell her we write numbers as groups of ten, and have her link the ten loose cubes into a row. Write the number 20, and ask her what the first digit means (we have 2 tens). Ask her what the second digit, 0, means that there are no more than 2 groups of 10.

Show your student the number-word cards eleven through twenty, or write the number words down, and help him to read them. Ask him to match them with the number cards 11-20. Continue to provide practice in reading and writing the numbers as desired or needed.

Discussion

Pages 62-63

You can have your student use dot cards and match the number of dots to the numeral and the number word, as pictured here. In Singapore, egg trays are for 10 eggs. In the U.S., you can remove two cups from an egg carton and use that to represent a ten.

Task 1, p. 64

Have your student write the numbers as well as say them.

Workbook

Exercises 1-2, pp. 89-93

Reinforcement

Say a number between 10 and 20 and ask your student to write the number and tell you the numbers as tens and ones. For example, you say "thirteen" and your student writes '13' and says "ten and three."

(2) Relate numbers from 11-19 to a ten and ones

Textbook

Tasks 2-4, pp. 65-66

2. (a) fifteen
   (b) 15
   (c) 15
3. (a) fourteen
   (b) 14
   (c) 14
4. (a) 15   (b) 8, 18

Workbook

Exercise 3, pp. 94-95

1. (a) 16
   (b) 12
   (c) 19
   (d) 15
2. (a) 14
   (b) 18
   (c) 13
   (d) 17

Activity

Draw a blank number bond. Give your student a set of objects less than 20 and have him put them in a group of ten and a group of left-overs, and then write the numbers in the number bond and as an addition equation. Repeat with other examples.

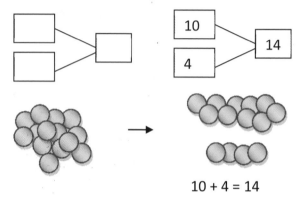

10 + 4 = 14

Discussion

Tasks 2-4, pp. 65-66

Workbook

Exercise 3, pp. 94-95

Reinforcement

If needed, continue to ask your student to practice addition and subtraction facts through 10. Add in the following facts:

10 + 1	10 + 2	10 + 3	10 + 4	10 + 5
10 + 6	10 + 7	10 + 8	10 + 9	
1 + 10	2 + 10	3 + 10	4 + 10	5 + 10
6 + 10	7 + 10	8 + 10	9 + 10	10 + 10

(3) Count back from 20, fill in missing numbers in a sequence

Activity

Mix up a set of numeral cards 0-20 and ask your student to arrange them in order. Then have her read them from 0 to 20, and then read them backwards from 20 to 0.

Set out 20 counters. Ask your student to count them by putting them into two rows of 10. Then have her count backwards by setting one counter to the side at a time.

Set out two numeral cards that are 3 or more apart, such as 13 and 18. Ask your student to place the correct numeral cards between them: 13, 14, 15, 16, 17, 18.

Repeat, but have your student fill in the missing numbers in reverse order. For example, set out the cards 14 and 9, and ask him to put the cards 13, 12, 11, and 10 between them.

Discussion

Tasks 5-6, pp. 66-67

Workbook

Exercise 4, pp. 96-97

Reinforcement

Use a set of number cards 1-20. Shuffle and place them face down. Turn over the top card and ask your student to say the numbers backwards starting with that number. Continue with the rest of the cards in the stack.

See if your student can see a pattern that skips one (odd or even numbers). Write the following patterns and ask your student to continue them.

➤ 2, 4, 6, 8, …

➤ 3, 5, 7, 9,…

➤ 20, 18, 16, 14,…

➤ 19, 17, 15, …

Textbook

Tasks 5-6, pp. 66-67 ✓

  6. (a) 16; 20
     (b) 10; 11
     (c) 15; 14; 12
     (d) 8; 14
     (e) 11; 9; 7

Workbook

Exercise 4, pp. 96-97

  1. picture of horse's head

  2. (a) 8
     (b) 11
     (c) 14
     (d) 18

(4) Compare and order numbers within 20

Textbook

Tasks 7-11, pp. 68-69

7. B

8. A

9. (a) B
    (b) C

10. (a) 17
    (b) 12

11. (a) 20
    (b) 8
    (c) 8, 12, 16, 20

Workbook

Exercise 5, pp. 98-100

1. (a) 8   (b) 9   (c) 4
    (d) 14   (e) 19   (f) 15

2. (a) 9   (b) 5   (c) 2
    (d) 9   (e) 8   (f) 18

3. (a) 7
    (b) 18

4. (a) 6
    (b) 14
    (c) 9

5. (a) 2, 3, 4, 5, 6, 7, 8
    (b) 20, 19, 18, 17, 16,
         15, 14, 13

6. (a) 10, 12, 13, 15, 20
    (b) 18, 15, 14, 11, 9

Activity

Set out two sets of counters with between 11 and 20 in each set. Ask your student to arrange them in tens and ones, write the number for each set, and tell you which one of the sets has the greater number and which has the smaller number.

Give your student two sets of dot cards representing two numbers between 11 and 20, such as 17 and 15, and ask your student to write the numbers for them and tell you which is the greater number and which is the smaller number. Lead her to see that if both numbers have tens, she just needs to compare the ones.

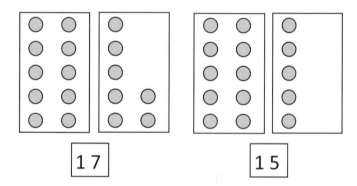

Repeat with three sets (3 numbers). Ask your student which is the greatest number and which is the smallest number. Do another example with 3 sets of dot cards, but this time include a 1-digit number.

Discussion

Tasks 7-11, pp. 68-69

Your student can probably answer Tasks 7-9 without even counting the items. You can have her count them and write down the numbers on a whiteboard or separate sheet of paper, and then after finishing the tasks in the book have her compare them without the aid of a picture. Then have her compare other numbers.

Activity

Give your student 4 to 5 numeral cards, such as 9, 16, 20, and 11 and ask her to put them in order.

Ask your student some questions such as:

➢ What number is between 12 and 14?

➢ What number comes before 15?

➢ What number comes after 16?

Reinforcement

Extra Practice, Unit 6, Exercise 1, pp. 55-58

Game

Material: 2-4 sets of number cards 0-20.

Procedure: Shuffle and deal all out. Each player keeps his or her cards face down. Each player turns over the top card. The player with the greatest number gets all the cards. If two players have the same number, they each turn over another card to see who wins the round.

Tests

Tests, Unit 6, 1A and 1B, pp. 69-74

Chapter 2 – Addition and Subtraction

Objectives

- Add and subtract within 20 using various strategies.
- Practice and memorize basic addition and subtraction facts within 20.

Material

- Two ten-cup egg cartons (remove the extra two cups from a regular egg carton)
- Objects that fit into the egg cartons
- 10-frame (5 by 2 array of squares) and counters
- Dot cards 1-20
- Counters
- Multilink cubes
- Playing cards 1-10
- Hundreds board
- Two number cubes, labeled 5-10
- Number cards 1-20
- Cards with '+', '−', or '=' on them (operation cards)

Notes

Students should have addition and subtraction facts through 10 memorized before beginning this unit.

The following thinking strategies for addition are introduced:

> Make 10: Split one of the numbers up in order to make a ten with the other number (pp. 62-64)

$$8 + 5 = 10 + 3 = 13$$
$$\underset{2\quad 3}{\diagup\diagdown}$$

> Add the ones: To add ones to a ten and ones, add the ones (p. 65). Numbers where the sum is greater than 20 will be dealt with in *Primary Mathematics 1B*.

$$13 + 5 = 10 + 8 = 18$$
$$\underset{10\quad 3}{\diagup\diagdown}$$

> Subtract from the ones: To subtract ones from a ten and ones when there are enough ones, subtract the ones from the ones (pp. 65-66).

$$18 - 5 = 10 + 3 = 13$$
$$\underset{10\quad 8}{\diagup\diagdown}$$

> Subtract from the ten: To subtract ones from a ten and ones where there are not enough ones, subtract the ones from the tens and add that result to the ones (p. 66).

$$13 - 8 = 3 + 2 = 5$$
$$\underset{3\quad 10}{\diagup\diagdown}$$

➤ Count on: When one of the numbers is 1, 2, or 3, count on from the other number to find the answer (p. 67).

➤ Count back: When the number being subtracted is 1, 2, or 3, count back from the other number to find the answer (p. 67).

Your student should practice these strategies and should work out number facts using them before memorizing the facts. Provide adequate concrete illustrations, particularly with the strategies of making ten and subtracting from a ten. This concept will be used many times throughout the series in a variety of contexts, and will contribute to mental math strategies for larger numbers.

After lesson 1 add the following addition facts to your fact practice and provide daily, short practice in math facts. Do not expect your student to have instant recall on these facts, but eventually she should be able to find the answers quickly, "making a ten" mentally if needed.

9 + 2	9 + 3	9 + 4	9 + 5	9 + 6	9 + 7	9 + 8	9 + 9
	8 + 3	8 + 4	8 + 5	8 + 6	8 + 7	8 + 8	8 + 9
		7 + 4	7 + 5	7 + 6	7 + 7	7 + 8	7 + 9
			6 + 5	6 + 6	6 + 7	6 + 8	6 + 9
				5 + 6	5 + 7	5 + 8	5 + 9
					4 + 7	4 + 8	4 + 9
						3 + 8	3 + 9
							2 + 9

After lesson 4, add the following subtraction facts to your daily fact practice.

11 – 10	11 – 9	11 – 8	11 – 7	11 – 6	11 – 5	11 – 4	11 – 3	11 – 2
12 – 10	12 – 9	12 – 8	12 – 7	12 – 6	12 – 5	12 – 4	12 – 3	
13 – 10	13 – 9	13 – 8	13 – 7	13 – 6	13 – 5	13 – 4		
14 – 10	14 – 9	14 – 8	14 – 7	14 – 6	14 – 5			
15 – 10	15 – 9	15 – 8	15 – 7	15 – 6				
16 – 10	16 – 9	16 – 8	16 – 7					
17 – 10	17 – 9	17 – 8						
18 – 10	18 – 9							
19 – 10								

(1) Add two 1-digit numbers whose sum is greater than 10 by making a 10

Textbook

Pages 70-71

Tasks 1-2, pp. 71-72

1. (a) 14
   (b) 15

2. (a) 10   17
   (b) 10   14
   (c) 10   12
   (d) 10   13
   (e) 10   13

Workbook

Exercise 6, pp. 101-102

1. 13   18
   16   17
   12   15
   14   11

2. 12   16
   13   13
   11   13
   11   15

Exercise 7, pp. 103-104

1. 12   14
   15   12
   14   11
   14   12

2. 13   11
   11   15
   13   18
   11   11

Activity

Use two 10-cup egg cartons and objects that fit into the cups as a 10-frame, or draw two 5 by 2 arrays of squares so that each square holds a counter. You can also mark off the arrays on the blank squares on the back of some laminated hundred boards. Egg cartons have the advantage of keeping the objects in place.

Put 9 counters (objects) on one 10-frame (in the cups of the egg cartons) and 1 on the other. Ask your student for the total, and move 1 counter over to make a ten. Say: 9 plus 1 is 10. Then put 9 counters on one frame and 2 on the other. Ask your student for the total, and move 1 counter over to make a ten. Say: 9 plus 2 is 10 and 1, or 11.

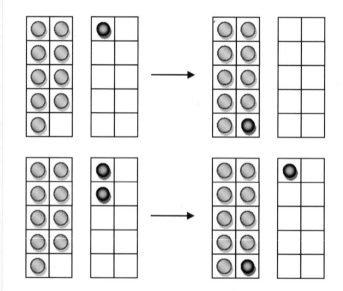

Repeat with 9 + 3. Ask your student to move counters to show other addition problems where the sum is greater than 10, such as 7 + 8 or 6 + 9. Point out that the discs do not have to be moved to the first 10-frame; in 7 + 8 they can make a ten with 8.

Then, do the same thing, but this time show the process with number bonds and write the addition equation.

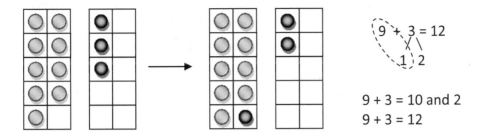

9 + 3 = 10 and 2
9 + 3 = 12

Discussion

Pages 70-71

Tasks 1-2, pp. 71-72

Allow your student to use manipulatives for task 2, if necessary. She should do task 2 across, rather than down, to see the relationship between the two problems in each row; the first problem in each row makes a ten, and the second more than a ten.

Workbook

Exercises 6-7, pp. 101-104

Reinforcement

The exercises include pictures and your student could simply be counting to find the total. Rather than going on immediately to the next section in the text, spend some time working with your student on the concept of "making a ten" until she can solve all problems involving addition of two single digit numbers where the sum is within 20 without constant use of manipulatives. She does not need to achieve any speed yet, just be comfortable with computing the sums by making a ten mentally. Use the following activities, or ones of your own devising.

Use 10 multilink cubes stuck together in a rod, and 9 individual cubes, or a 10-rod and 9 unit cubes from a base-10 set, or any other base-10 representation, such as 10 craft sticks with a rubber band around them and 9 singles. Do not let your student break the ten down into units in the activity that follows. If she understands that a dime is equivalent to 10 pennies, you can use 1 dime and 9 pennies.

Give your student 6 to 9 units. Then tell him a number between 5 and 9 and ask him to show you the total of the two numbers with the objects. Write the equation, e.g. 6 + 8. Since he only has a total of 9 units, he will have to use the ten to show the answer. For example, if you give him 6 units and ask him to show you 6 + 8, he will have to trade in 2 units for the ten to show 4 and 10 as the answer. Or, if you give him 9 units and ask him to show you 9 + 7, he will have to trade in 3 units for the ten to show you 10 and 6 as the answer.

Write some problems in the following format, and ask your student to fill in the blanks:

➢ 9 + 5 = 10 + _____

➢ 4 + 7 = 10 + _____

If your student has trouble, discuss the number bond that can be formed with one of the numbers to make a ten with the other number.

As your student gets more proficient on computing the sum mentally, you can start working for speed as you continue with the lessons, providing math fact practice each day.

You can begin to have your student memorize doubles. He should already know:

$$1 + 1 = 2 \qquad\qquad 2 + 2 = 4 \qquad\qquad 3 + 3 = 6$$

$$4 + 4 = 8 \qquad\qquad 5 + 5 = 10$$

Have your student learn the following doubles:

$$6 + 6 = 12 \qquad\qquad 7 + 7 = 14$$

$$8 + 8 = 16 \qquad\qquad 9 + 9 = 18$$

Then, you can use dot cards to show another strategy for adding two numbers that differ by 1, or "doubles + 1."

$$6 + 7 = 12 + 1 = 13$$

$$7 + 8 = 14 + 1 = 15$$

$$8 + 9 = 16 + 1 = 17$$

Your student can use either method (doubles + 1 or making a ten) to find the answers to these.

Mental Math 12-14

Games

Material: 4 sets of number cards, 1-10.

Procedure: Shuffle and deal all cards out. Each player turns over two cards and adds the numbers on their card. The player with the highest total gets all the cards. After all cards have been turned over, the player with the most cards wins.

Game

Material: Two number cubes labeled 5-10, counters, a game-board with the numbers 10-20 written randomly on a grid (You can use as large a grid as you like. For example, use a wet-erase marker and write the numbers 10 to 20 randomly in the blank squares on the back of a laminated hundred-chart.)

Procedure: Players take turns throwing the number cubes, adding the numbers, and putting their counter on the answer on the game board. The first player to get three counters in a line (vertically, horizontally, or diagonally) wins.

(2) Add a 1-digit number to a 2-digit number within 20 by adding ones

Activity

Use two 10-frames and counters. Write an addition expression where a 1-digit number is added to a 2-digit number with a sum within 20, such as 12 + 6. Have him place 12 counters on the ten-frame and 6 by themselves. Then have him add the 6 to the ten-frame and tell you the answer to the expression. Repeat with some other examples. Then, do the same thing but show the process with a number bond and write the equation. Point out that we can think of the 2-digit number as ten and ones, and add the ones together. We will then have the ten and the total ones for the answer.

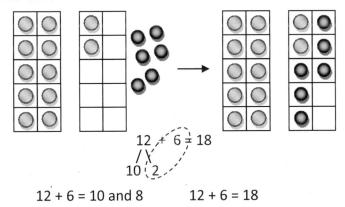

12 + 6 = 18

10   2

12 + 6 = 10 and 8   12 + 6 = 18

Do a few problems where adding the ones make another ten, such as 13 + 7. Ask your student to add the tens, and then realize that she now has 2 tens, so the answer is 20.

13 + 7 = 20   13 + 7 = 10 and 10

10   3   13 + 7 = 20

Discussion

Tasks 3-4, p. 73

Workbook

Exercise 8, pp. 105-107

Reinforcement

Mental Math 15

Extra Practice, Unit 6, Exercise 2A, pp. 59-60

Textbook

Tasks 3-4, p. 73

3. 17

4. (a) 19   19
    (b) 20   20

Workbook

Exercise 8, pp. 105-107

1. 15   19
    17   18
    19   18
    20   19

2. 19   18
    20   16
    17   18
    17   17

3. 20   6; 17
    4; 16   5; 15
    8; 19   4; 14
    3; 17   5; 18

(3) Subtract a 1-digit number from a 2-digit number by subtracting ones

Textbook

Tasks 5-6, pp. 73-74

5. 12

6. (a) 15   (b) 12

Workbook

Exercise 9, pp. 108-110

1. 12   10
13   12
13   10
13   14

2. 12   11
12   17
12   11
13   15

3. 11   3; 14
2; 10   7; 10
3; 10   2; 12
10; 10   4; 12

Activity

Use two 10-frames and counters, as in the previous activities, or a 10-rod of multilink cubes and ones.

Write a subtraction expression where a 1-digit number is subtracted from a 2-digit number where the ones is greater than the ones being subtracted, such as 18 – 6. Illustrate the first number with the 10-frames and counters, or multilink cubes. Ask your student to take away the second number. It should be obvious that she needs to simply take away the correct number of ones. Illustrate a few other examples, and then include a drawing of a number bond, showing the first number as 10 and ones, and the second number being subtracted from the ones.

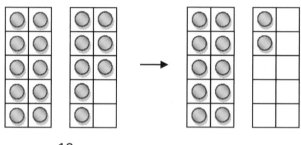

18 ⟨ 10 / 8 – 6 = 2

18 – 6 = 10 and 2
18 – 6 = 12

Write some similar subtraction problems and see if your student can find the answer. Draw a number bond if needed. Only use subtraction situations where there are enough ones to take away from. Include an example where all the ones are subtracted, e.g. 13 – 3. In each case, point out that there are enough ones to subtract from.

19 – 5 = 14
/ \
10   9

17 – 3 = 14
/ \
10   7

13 – 3 = 10
/ \
10   3

15 – 5 = 10
/ \
10   5

Only use subtraction situations where there are enough ones to take away from. Include an example where all the ones are subtracted, e.g. 13 – 3.

Write the expression 20 – 10 and ask your student to solve it. There are 2 tens, take away 1 ten, that leaves 1 ten. 20 – 10 = 10.

Discussion

Tasks 5-6, pp. 73-74

Reinforcement

Mental Math 16

WK BK 112 & 113

11–13
Finish Excer. Text
pgs
75-78

then
WK BK
114–121

(4) Subtract from a 2-digit number by subtracting from a ten

Textbook

Tasks 7-8, p. 74

7. 8

8. (a) 5   (b) 4
       6   5
       9   8

Workbook

Exercise 10, pp. 111-113

1. 5   6
   4   2
   5   7
  10   2

2. 6   5
   8   2
   4   8
   6   7

3. 9   7
   7   9
   8   8
   7   9

Activity

Use counters and two 10-frames.

Write a subtraction expression where a 1-digit number is subtracted from a 2-digit number when the ones is smaller than the ones being subtracted, e.g. 15 – 7. Show the first number on the ten-frames. Tell your student that he can solve this problem using facts he already knows. 10 – 7 = 3. Remove 7 from the full 10-frame. This leaves 3. Ask your student for the total amount left. There are 3 + 5 = 8 left. So, 15 – 7 = 8.

15 – 7 = 8

Now, show the same process with 13 – 6, but this time also draw a number bond to show that we are subtracting from a ten.

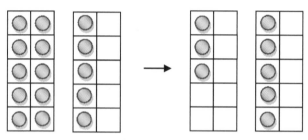

$$10 - 6 = 4$$
$$13$$
$$3$$

13 – 6 = 4 and 3
13 – 6 = 7

Repeat with other examples, eventually just drawing the number bond and not using manipulatives.

Remind your student that in subtraction, we are finding a missing part.

Write the equations;

➢ $13 - 6 = $ ____

➢ $6 + $ ___ $= 13$

Tell your student that she can find the missing part by thinking of how many more are needed to make 10 with 6 (4) and then add on the remaining 3. (4 + 3 = 7)

Write the expression $20 - 10$ and ask your student to solve it. There are 2 tens, take away 1 ten, that leaves 1 ten. $20 - 10 = 10$.

Write some problems involving subtracting from 20, e.g. $20 - 4$. Show that we can subtract from a ten. If we know that $10 - 4 = 6$, then we know that $20 - 4 = 10$ and 6, or 16.

$$10 - 4 = 6$$
$$20 \diagdown$$
$$10$$

$$20 - 4 = 10 \text{ and } 6$$
$$20 - 4 = 16$$

Give your student a 10-rod of multilink cubes and 9 unit cubes. Tell him the 10-rod cannot come apart. Ask him to show you the answer to some problems such as $14 - 8$. He first shows 14 with a ten and 4 ones. Then he must take away 8. To do this he will have to trade in the 10-rod for 2 ones.

Discussion

Tasks 7-8, p. 74

Allow your student to use a 10-rod made from the multilink cubes if needed.

Workbook

Exercise 10, pp. 111-113

Reinforcement

Provide plenty of additional practice in subtraction problems involving taking away from the ten.

If your student has a lot of difficulty with the concept, or with going from using manipulatives to finding the answer mentally, you can teach another strategy:

$13 - 6$: First take away 3 from the ten. That is not taking away enough. We need to take away another 3 (since $6 - 3 = 3$). Essentially, we are doing the subtraction in two steps:

$13 - 6$: $13 - 3 = 10$, $10 - 3 = 7$

Repeat with some other examples, such as:

17 – 9: Take away 7 to get down to 10, need to take away another 2 to take away 9 in all, the answer is 8.

Write some problems in the following format, and ask your student to fill in the blanks:

➢ 13 – 6 = _____ + 3

➢ 14 – 7 = 3 + _____

If your student understands that a dime is worth the same as 10 pennies, you can illustrate subtracting from a ten as making change. Give him a dime and 5 pennies. Say, "You have a dime and 5 pennies. You want to buy something that costs 7 pennies. How can you pay for it?" Guide her through the following: 5 pennies are not enough, so she have to pay with the dime. She gets 3 pennies change. 3 pennies from the change and 5 pennies she still has is 8 pennies. 15 – 7 = 8.

Ask your student to make change for other similar situations.

Extra Practice, Unit 6, Exercise 2B, pp. 61-62

Math fact practice

Use 4 sets of number cards 1-10. Shuffle the cards and place all face-down. Select a target number, such as 14, and write it down. Turn over the cards one at a time and ask your student to subtract the card from the target number and give the answer. After a few problems change the target number.

Mental Math 17-19

(5) Add or subtract 1, 2, or 3 by counting on or back

Activity

Write the numbers 1 to 20 in a row, or ask your student to lay out the number cards 1 to 20 in a row.

Point to a number and ask your student to add or subtract 1, 2, or 3 from it by counting on or back. For example, point to 11 and say, "Subtract 3." Ask her to count back: 10, 9, 8. Make sure she does not use the number that you are pointing at as one of the numbers that is being added or subtracted.

Then, ask your student to do the same thing without a number line. Write down a number within 20 and tell him to add or subtract 1, 2, or 3. For example, write down 11 and tell him to add 2.

Discussion

Tasks 9-12, p. 75

Workbook

Exercise 11, pp. 114-115

Reinforcement

Continue to ask your student to practice addition and subtraction, using fact cards, games, and activities from previous lessons.

Mental Math 20

Textbook

Tasks 9-12, p. 75

9. 11

10. (a) 11   (b) 12
    (c) 16   (d) 19

11. 9

12. (a) 8   (b) 9
    (c) 13   (d) 16

Workbook

Exercise 11, pp. 114-115

1. 11   15   14
    13   18   18
    20   14   12
    14   20   20
    19   16   16
    12   19   13

    15   16   8
    19   13   17
    11   19   14
    17   14   9
    13   17   18
    10   11   11

2. 18, 19, 17, 17, 16
    18, 19, 19,
    17, 17, 16, 18,
    19, 19, 17, 19, 19, 18
    20

(6) Add and subtract within 20

Textbook

Tasks 13-17, pp. 76-78

14. (a) 12   (b) 12   (c) 13
    (d) 15   (e) 16   (f) 19

16. (a) 8   (b) 9   (c) 6
    (d) 8   (e) 6   (f) 10

17. (a) 14   (b) 15
    (c) 13   (d) 15
    (e) 12   (f) 17
    (g) 13   (h) 12
    (i) 13   (j) 16
    (k) 14   (l) 13
    (m) 11   (n) 12
    (o) 8   (p) 7
    (q) 9   (r) 8
    (s) 6   (t) 3
    (u) 4   (v) 7
    (w) 7   (x) 7
    (y) 9   (z) 5

Workbook

Exercise 12, pp. 116-118

1. 12   15
   8   5
   7   16
   11   3

2. 11   6   8
   12   6   8
   6   12   9
   12   13

3. Check answer.

Activity

Show your student two sets of counters or multilink cubes, using different colors. For example, set out 5 blue counters and 7 red ones. Guide your student in writing two addition and two related subtraction problems for the situation.

$7 + 5 = 12$ $5 + 7 = 12$
$12 - 7 = 5$ $12 - 5 = 7$

Write some addition and subtraction equations without the plus or minus sign, and ask your student to fill in the symbols. For example:

➢   18 ◯ 5 = 13
➢   5 ◯ 6 = 11

Give your student some number cards and cards with the plus, minus, and equal sign on them and ask him to form an equation.

Tell your student some simple word problems involving addition or subtraction within 20 and have her write the expression that would be used to solve the problem and find the answer.

Discussion

Tasks 13-17, pp. 76-78

Tasks 13 and 15 illustrate some games that you can play with your student for practicing math facts. You can try these games or use other means for practicing the math facts. Continue with daily math fact practice as you continue with the remaining units.

Workbook

Exercises 12-15, pp. 116-123

You can save some of these to do later as review after working on the math facts as you continue with the next units.

For Exercise 14, your student might not even need to find the answers to the expressions if he has good number sense and can take

Reinforcement

Mental Math 21-22

Write the following for your student to solve:

➢ $13 + 7 = 5 + \square$

➢ $15 - 8 = 17 - \square$

➢ $9 + 5 = 20 - \square$

➢ $14 \bigcirc 3 = 10 \bigcirc 1$

➢ $18 \bigcirc 2 = 15 \bigcirc 5$

Extra Practice, Unit 6, Exercise 2C-2D, pp. 63-68

Tests

Tests, Unit 6, 2A and 2B, pp. 75-81

Workbook

Exercise 13, pp. 119-121

1. −   +
    +   −
    +   +
    −   −
    −   +

2. Answers may vary.
    $8 + 4 = 12$
    $9 + 5 = 14$
    $16 - 9 = 7$
    $13 - 5 = 8$

3. $7 + 6 = 13$ $13 - 6 = 7$
    $6 + 7 = 13$ $13 - 7 = 6$
    $11 + 5 = 16$ $16 - 5 = 11$
    $5 + 11 = 16$ $16 - 11 = 5$

4. $18 - 13 = 5$ $13 - 5 = 8$
    $3 + 6 = 9$ $9 + 0 = 9$
    $11 + 9 = 20$ $9 + 9 = 18$
    $16 - 10 = 6$ $20 - 17 = 3$
    $4 + 6 = 10$

Exercise 14, p. 122

1. (a) $6 + 6 \longleftrightarrow 5 + 7$
    (b) $9 + 7 \longleftrightarrow 8 + 8$
    (c) $8 + 3 \longleftrightarrow 5 + 6$
    (d) $7 + 6 \longleftrightarrow 9 + 4$
    (e) $9 + 6 \longleftrightarrow 8 + 7$
    (f) $7 + 7 \longleftrightarrow 6 + 8$
    (g) $8 + 9 \longleftrightarrow 9 + 8$

Exercise 15, p. 123

1. (a) 10   (b) 10
    (c) 8   (d) 7
    (e) 8   (f) 7
    (g) 9   (h) 2
    (i) 4   (j) 6
    (k) 8   (l) 6

Review

Workbook

Review 4, pp. 124-127

1. (a) 9   (b) 8
   (c) 11   (d) 12
   (e) 14   (f) 18
   (g) 20   (h) 17
2. $6 + 7 = 10 + 3$ $9 + 3 = 10 + 2$
   $8 + 8 = 10 + 6$ $7 + 7 = 10 + 4$
3. $8 + 6 = 14$ $14 - 8 = 6$
   $6 + 8 = 14$ $14 - 6 = 8$
4. (a) 10, **11**, **12**, **13**, 14, **15**, **16**, **17**, **18**, 20
   (b) 20, **19**, 18, **17**, **16**, 15, **14**, **13**, **12**, **11**
5. (a) 19   (b) 9
6. (a) draw 12 marbles
   (b) 13
7. 1st   3rd   2nd   4th
8. H
9. (a) 10   (b) 8   (c) 0   (d) 10
10. $14 + 5 = 19$; 19
11. $12 - 4 = 8$; 8

Review 5, pp. 128-131

1. (a) 15   (b) 20
2. (a) 10   (b) 5   (c) 10   (d) 2
3. (a) 12   (b) 19   (c) 14   (d) 16
4. (a) B   (b) C
5. (a) 13   (b) 4
6. (a) 10   (b) 2
        11   3
        12   4
7. (a) 7   (b) 11
8. (a) 19; 15; 0; 6
   (b) 6; 4; 10; 14
9. $12 - 5 = 7$; 7
10. $14 + 3 = 17$; 17
11. $15 - 8 = 7$; 7

Note

The workbook has 2 review exercises in succession. You can use them as assessments to see if you need to go back and re-teach any part of units 1-6, or save them and ask your student to do several problems a day as you go on as a form of continuous review.

Reinforcement

Mental Math 23-24

Tests

Tests, Cumulative Test Units, 1-6, A and B, pp. 83-88

Unit 7 – Shapes

Chapter 1 – Common Shapes

Objectives

- Recognize and name the four basic shapes: circle, triangle, square, and rectangle.
- Sort and classify objects by shape, size, color, or orientation.
- Describe or continue a pattern according to one or two attributes such as shape, size, or color.
- Choose suitable shapes to fit together to make a basic shape.

Material

- 3-dimensional models of cubes, rectangular and triangular prisms, cylinders, cones, and pyramids
- Attribute blocks or cutouts of 3 different sizes each of rectangles, squares, circles, and triangles in 3 or 4 colors
- Pattern blocks or a set of 4-5 cutouts each of 4-5 different shapes
- Shape stickers, if available

Notes

The emphasis in this unit is on practical work. Students need to know the names of the shapes for circle, square, rectangle, and triangle. At this stage, you can have your student regard squares as different shapes than rectangles. In the *Primary Mathematics* curriculum students are not required to use the words cube, cone, sphere, cylinder, etc. It is up to you if you want to teach your student these terms. Use real objects rather than 2-dimensional drawings of 3-dimensional shapes.

Students will be sorting objects by shape, size, or color. Sorting objects by shapes requires the student to pay attention to number of sides and corners.

Students will be identifying and continuing patterns based on shape, size, color, or orientation. Some students can see patterns, such as those on p. 75 of the text, more easily when they verbalize them. Ask your student to "say" the pattern, e.g. rectangle, circle, triangle, rectangle, circle, triangle, ... or upright, sideways, upright, ... At this stage only one or two attributes (out of shape, size, color, or position) change. For example, the pattern in (d) changes by position and color, but shape and size stays the same.

(1) Identify common shapes

Textbook

Page 79

Tasks 1-4, pp. 80-83

4. (a) 4 corners, 4 sides
   (b) 5 corners, 5 sides
   (c) 6 corners, 6 sides

Workbook

Exercise 1, pp. 132-134

1. Check answers.
2. Shape that should be colored:
   Third; Third; Second
   First; Second
3. rectangle, circle, triangle, circle,
   rectangle

Exercise 2, pp. 135-137

1. triangle
   circle,
   rectangle
   rectangle

2. (a) square   (b) circle
   (c) circle   (d) triangle
   (e) square   (f) rectangle
   (g) triangle   (h) triangle

Exercise 3, pp. 138-141

1. (b), (c), (e)
2. (c), (d), (e)
3. (a), (d), (d)
4. (b), (c), (g), (h)
5. (a) 4, 4   (b) 4, 4
   (c) 8, 8   (d) 8, 8
   (e) 10, 10

Activity

Give your student some 3-dimensional models and discuss their shapes and characteristics, e.g., number of faces, edges, and corners, whether they roll or not, whether they can be stacked or not. Extend the discussion to similar objects around the room, e.g. tin cans, rolling pin, balls, boxes, dice, etc.

Draw a circle, triangle, square, and rectangle. Label each and make sure your student knows the names of these four shapes. Ask her to describe attributes of the shapes. Some shapes have no corners; others have sharp corners and flat sides. Ask her to count the number of corners and sides of these four shapes. Tell her that a square is just like a rectangle, but all its sides are equal in length.

Discussion

Page 79

Ask your student to talk about the different types of building blocks and shapes that children in the picture could trace from the faces. Then ask him to trace the faces of the 3-dimensional models onto paper. Ask him to name the faces he has traced on the paper.

Tasks 1-4, pp. 80-83

Workbook

Exercises 1-3, pp. 132-141

Reinforcement

Ask your student to draw a picture consisting of the various shapes put together, such as a picture of a robot.

Extra Practice, Unit 7, Exercise 1A, pp. 73-76

Activity

Give your student some attribute blocks or cutouts of different sizes, shapes, and colors of circles, rectangles, squares, and triangles and ask her to sort them in different ways. She could sort by shape, by size, by number of edges, or by color. You can discuss sub-groups as well. For example, she could first sort the shapes into those with no corners or those with corners, then sort the shapes with corners into shapes with 3 corners and shapes with 4 corners, and then sort the shapes with 4 corners into rectangles and squares. You can draw a tree diagram of the different categories, for example:

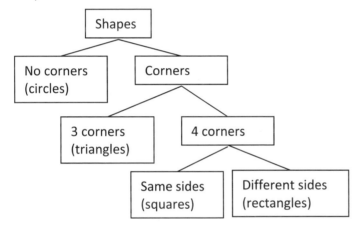

Discussion

Task 5, pp. 84-86

Workbook

Exercises 4-5, pp. 142-146

Extension

Ask your student to sort shapes that are not similar, such as different types of triangles (right, acute, or obtuse triangles) and different shapes of rectangles (tall and thin, short and fat, etc.).

Discuss and sort other shapes, such as ovals, parallelograms, trapezoids, or hexagons.

Textbook

Task 5, pp. 84-86

Workbook

Exercise 4, pp. 142-144

Check answers.

Exercise 5, pp. 145-146

Check answers.

(3) Identify patterns

Textbook

Tasks 6-7, pp. 87-88

3. (a) same shape, different size
   (b) different shape, different size
   (c) same shape and size
   (d) same shape and size

4. (a) triangle
   (b) small red triangle
   (c) a rectangle on its side
   (d) blue upside triangle pointing down

Workbook

Exercise 6, pp. 147-148

1. (a) circle
   (b) rectangle
   (c) smaller square
   (d) rectangle
   (e) triangle

2. row 2: square, rectangle
   row 3: rectangle, circle, triangle, square
   row 4: triangle, rectangle

Activity

Draw or use cut-out shapes to show two shapes that have the same shape and color but different size. Ask your student to identify similarities and differences between them.

Repeat with other pairs of shapes that differ by one attribute, such as same shape and size but different color. Include examples where the pair only differs in orientation.

Now show pairs of shapes where two of the attributes are different, such as size and shape.

Draw or use cut-out shapes or pattern blocks to display a sequence of shapes. Ask your student to read the pattern out loud to determine what comes next in the pattern. For example:

Triangle, triangle, square, triangle, triangle, square...

Ask your student to draw boxes or circles around the repeated patterns.

Discussion

Task 6-7, pp. 87-88

For Task 7 get your student to read the pattern aloud and find out what comes next.

Workbook

Exercise 6, pp. 147-148

Reinforcement

Ask your student to look for patterns in the environment.

Ask your student to create patterns with stickers or pattern blocks.

Extra Practice, Unit 7, Exercise 1B, pp. 77-78

(4) Combine shapes

Textbook

Tasks 8-10, pp. 89-90

5. Pink with red, orange with blue, purple with green

6. Blue with green, pink with purple

Activity

Before the lesson put some pattern blocks or cut-out shapes together and trace the outline. Then ask your student to use a selection of shapes and try to find out which shapes were used to make the outline.

Cut some basic shapes into two parts in various ways, mix them up, and ask your student to try to pair them up to make the original shapes (see p. 76 of the textbook).

Discussion

Tasks 8-10, pp. 89-90

Workbook

Exercise 7, p. 149

Reinforcement

Ask your student to put a jigsaw puzzle together.

Tests

Tests, Unit 7, 1A and 1B, pp. 89-96

Tests, Cumulative Test Units 1-7, A and B, pp. 97-104

Workbook

Exercise 7, p. 149

Check answers.

Unit 8 – Length

Chapter 1 – Comparing Length

Objectives

- Compare the length of two or more objects by direct comparison or by counting units.
- Compare the length of two or more objects by indirect comparison.
- Arrange objects in order by length.

Material

- Multilink cubes
- Paper or cardboard strips

Notes

Students should already be familiar with terms such as tall, short, and long, and with terms dealing with comparison such as shorter than and longer than. They should know the difference between size and length, and between length and height.

Students will compare length in a variety of ways:

- ➤ Physical appearance: One object is obviously longer than another (textbook pp. 91-92).

- ➤ Direct comparison: Place the objects side by side and have the same starting point (this is actually an important concept in model drawing, which will come later).

- ➤ Counting identical units: Compare the length by comparing the number of units (textbook pp. 93-94).

- ➤ Measure using non-standard units: Place a set of identical units end to end to match the length of each object, and then count the number of units to compare (textbook pp. 95-96).

- ➤ Indirect comparison: Cut a paper strip, ribbon, or string to represent the length of one object, or mark its length on a straight edge, and compare it with the length of another object, or cut a paper strip for each object and compare them.

Give your student concrete experience with each of these methods – not just look at the pictures in the textbook.

(1) Compare length

Textbook

Pages 91-92

Tasks 1-4, pp. 93-94

1. The green string is longest.
   The yellow string is shortest.

2. The purple stack of blocks is tallest.
   The yellow stack of blocks is shortest.

3. Worm C takes the longest path.
   Worm B takes the shortest path.

4. Tape R is the longest
   Tape Q is the shortest.

Workbook

Exercise 1, pp. 150-151

Check answers.

Exercise 2, pp. 152-153

1. (a) C
   (b) B
   (c) A
   (d) C

2. C; B; D; A

Activity

Put multilink cubes into rods of 4 lengths, 2 of which are the same, and ask your student to compare their lengths. Ask your student to line them up and say sentences such as:

"The red one is as long as the blue one."
"The yellow one is longer than the red one."
"The yellow one is the longest."
"The green one is the shortest."

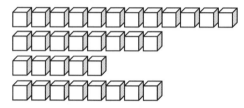

Then ask your student how she would compare the lengths if the rods were not placed side by side. Since the cubes are all of the same length, the lengths can be compared by counting the cubes. Have her count the cubes and compare the lengths.

Put two objects of different but similar length at different places in the room. Ask your student how he could find out which one is longer if he cannot move them next to each other. Allow him to try out his ideas. Give him a paper strip and lead him to see that he can use a third object that is longer than both, mark off the length of each on the strip, and compare the marks.

Discussion

Pages 91-92

Tasks 1-4, pp. 93-94

Workbook

Exercises 1-2, pp. 152-153

Reinforcement

Extra Practice, Unit 8, Exercise 2, pp. 83-84

Tests

Tests, Unit 8, 2A and 2B, pp. 109-112

Chapter 2 – Measuring Length

Objectives

- Estimate and measure length with non-standard units.

Material

- Multilink cubes
- Paper clips (2 different sizes)

Notes

In this section, students will measure lengths using non-standard units such as paper clips, multilink cubes, and craft sticks. The standard units (e.g. centimeter, meter, inch, foot, yard) will be introduced in *Primary Mathematics 2*.

To measure the object, the student will place the units end to end alongside the objects to match the length. They should see that it is not always possible to have an exact number of units to match the length. Use the word 'about' to express the result of an approximate measurement.

Encourage your student to estimate a length before measuring it.

(1) Estimate and measure length

Textbook

Page 95

Tasks 1-2, pp. 95-96

Activity

Ask your student to measure an object with paper clips. It may be easier to use a chain of paper clips; since the measurements are not expected to be precise the slight difference when the paper clips are linked rather than separate is not significant. Since the length will probably not end at exactly the end of a paper clip, discuss approximate lengths using the term "about." An object that is a bit more than 8 paper clips long is about 8 paper clips long, whereas one that is a bit less than 9 paper clips long is about 9 paper clips long. Write down the length, along with the unit, e.g. about 8 paper clips.

Then ask your student to measure the same object with a different unit, such as larger paper clips. Write down the length along with the unit, e.g. about 9 short paper clips and about 5 large paper clips.

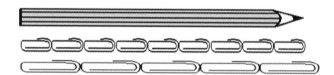

Workbook

Exercise 3, pp. 154-156

1. 7
   7
   5
   7

2. 5
   6
   12

3. middle one

4. (a) 7
   (b) 5
   (c) 3

Discuss the difference in numerical value for the lengths. The length of an object will have different values when it is measured with different units. Lead your student to see that an object's length will have a larger value when measured with a smaller unit.

Ask your student to measure the length of another object with the paper clips. Then ask him if the number of the length will be greater or smaller with the second type of unit.

Ask your student to estimate the length of a third object in both types of measuring units, and then measure their lengths.

Discussion

Page 95

Tasks 1-2, pp. 95-96

Workbook

Exercise 3, pp. 154-156

Reinforcement

Let your student compare the lengths or heights of various objects in the environment using chains of paper clips.

Start a growth chart on a wall to measure the student's height over time.

Measure the growth of a plant using non-standards units, such as paper clips. You can extend the paper-clip chain as the plant grows.

Extra Practice, Unit 8, Exercise 2, pp. 83-84

Tests

Tests, Unit 8, 2A and 2B, pp. 109-112

Unit 9 – Weight

Chapter 1 – Comparing Weight

Objectives

- Compare the weight of two or more objects by direct comparison or by counting identical units.

Material

- 3-4 objects of similar weight
- Simple balance
- Clay or Play-dough
- Multilink cubes
- Marbles, counters, or other objects that can be used as unit weights

Notes

Weight, like length and shape, is also an attribute of an object. Students are generally already familiar with the terms heavy, light, lighter, etc. used when comparing weight.

Weight can be compared in a variety of ways.

➢ Use the objects' heft: When two objects are obviously different we can compare them by heft or lifting. (Textbook p. 97)

➢ Direct comparison: Place objects on each side of a balance. The heavier one makes the balance on its side go down farther. (Textbook pp. 98)

➢ Measure using non-standard units: Use identical units to balance each object one at a time on the balance. Then compare the number of units needed. (Textbook p. 99)

➢ Indirect comparison: If one object weighs the same as a second object, we can measure the weight of a third object with the second object to compare its weight with the first object.

Again, give your students concrete experience. If you don't have a balance, you can build a simple one. Tape a ruler to the table so that it sticks out perpendicularly with a side. Hang a coat hanger on the ruler. Hang some paper cups from each side of the coat hanger. Encourage your student to estimate the weight when weighing with non-standard units.

Technically, weight is the measure of the gravitational pull between two objects, and mass is the measure of how much matter, or inertia, an object has. An object on the moon has the same mass as it has on earth, but it weighs less. When we use a balance to compare two objects, we are actually comparing their mass. When we use a scale, we are measuring the weight. Since the term weight is used more commonly for both, you can use that term for both weight and mass until your student learns the difference in science studies.

(1) Compare weight

Textbook

Pages 97-98

Tasks 1-3, pp. 98-99

1. (a) the ruler
   (b) the key

2. 5

Workbook

Exercise 1, pp. 157-158

1. (a) as heavy as
   (b) heavier than
   (c) lighter than

Activity

Ask your student to compare the weight of several objects of obviously different weights, using terms such as lighter, heavier, lightest, heaviest, as light as, as heavy as, same weight.

Use three or more objects that are similar in weight, such as a pear, an apple, and a banana, but different enough to register on whatever balance you are using, and ask your student how we could compare their weights. Show him a balance and have him use that to compare the weights of the objects and put them in order of weight. The objects have to be compared in pairs.

Ask your student how we could compare objects that we cannot put together on the balance, for some reason. Lead her to see that we can compare the weight to something that we could change to match, such as clay. Give her two objects and have her use clay to match the weight of one, and then compare the second object to the weight of the clay.

Lead your student to see that we could also use a set of identical units to balance each object, and then compare the number of units needed. Ask him to use multilink cubes to balance several objects, and then compare the weight of those objects by the number of multilink cubes needed.

Discussion

Pages 97-98, Tasks 103, pp. 98-99

Workbook

Exercise 1, pp. 157-158

Reinforcement

Young children often relate weight to size. Use some objects where the larger object is lighter than the smaller object and ask your student to compare their weights.

Extra Practice, Unit 9, Exercise 1, pp. 87-88

Tests

Tests, Unit 8, 2A and 2B, pp. 109-112

Chapter 2 – Measuring Weight

Objectives

- Estimate and measure weight with non-standard units.

Material

- 3-4 objects of similar weight
- Simple balance
- Multilink cubes
- Marbles, counters, or other objects that can be used as unit weights

Notes

In this chapter students will measure weight using non-standard units such as multilink cubes, counters, or marbles. The standard units (e.g. gram, kilogram, pound, ounce) will be introduced in *Primary Mathematics 2*.

Depending on the accuracy of your balance and the weight of the units, it may not always be possible to have an exact number of units to match the weight of an object. If one more object makes the balance dip more on the side with the units, you can say the weight of the object is "about" or "almost" the same as the number of units.

Encourage your student to estimate the weight of an object before measuring it.

(1) Estimate and measure weight

Textbook

Page 100

Tasks 1-2, p. 101

1. Box A = 6 units
   Box B =8 units
   Box C = 7 units
   Box B is heaviest.
   Box A is lightest.

Workbook

Exercise 2, pp. 159-160

1. (a) 7
   (b) 9
   (c) 3

2. (a) 5
   (b) 4

Exercise 3, pp. 161-162

1. (a) 7
   (b) 9
   (c) 8
   (d) watermelon
   (e) papaya
   (f) watermelon

2. (a) 6
   (b) 11
   (c) 5
   (d) C
   (e) C
   (f) B

Activity

Ask your student to measure the weight of an object with a balance using multilink cubes as the unit. Then ask her to weigh the same object with a different unit, such as counters. Write down the weights, along with the unit, e.g. 14 multilink cubes, 58 counters. Discuss the difference in numerical value for the weights. Ask your student to compare the weight of the units and lead her to see that it takes more of the lighter unit to weigh the object than of the heavier unit.

Ask your student to estimate the weight of other objects in various units, and then weigh them and record their weight in various units.

Discussion

Page 100

Tasks 1-2, p. 101

Workbook

Exercises 2-3, pp. 159-162

Reinforcement

Extra Practice, Unit 9, Exercise 2, pp. 89-90

Tests

Tests, Unit 9, 2A and 2B, pp. 117-121

Extension

Students do not learn about standard units of weight at this level, but many have experience with hearing that their weight is so many pound or kilograms. If you have pound or kilogram weights, you can introduce them informally, and ask your student to weigh various things using the pound or kilogram as the unit, and compare to their own weight. They could also weigh themselves and other things on a bathroom scale.

Unit 10 – Capacity

Chapter 1 – Comparing Capacity

Objectives

- Compare the capacity of two or more containers.

Material

- Several containers of different capacity
- Two smaller containers of different capacity

Notes

The capacity of a container is how much liquid it will hold. Investigation of capacity is a good way to give the student a 'feel' for volume.

In this section, students will learn various ways of comparing the capacity of two or more containers. Provide your student with concrete experience actually filling up and pouring out from various containers.

There are several ways to compare the capacity of two containers.

➢ Visually inspect containers. This is not a very accurate method, since the difference in capacity needs to be quite obvious, or the containers need to be similar. A shorter, wider container could have a greater capacity than a taller, narrower container, and your student might assume it has less capacity just because it is shorter. (Textbook p. 102).

➢ Fill up one container, and then pour the water from it into the other. If there is some remaining in the first when the second is full, the first has a greater capacity. If the first can be emptied into the second and not fill it, the second has greater capacity. (Task 1, textbook p. 103).

➢ Fill up each container and pour the contents into smaller containers of equal capacity to see how many more small containers one fills than the other (Task 3, textbook p. 104).

➢ Fill up both containers and pour the contents into two larger containers of equal size and shape and compare the levels.

(1) Compare capacity

Textbook

Page 102

Tasks 1-3, pp. 102-105

1. A

2. (a) B
   (b) C
   (c) C

3. (a) mug
   (b) 12
        8
        bottle
        cup

Workbook

Exercise 1, pp. 163-165

1. (a) glass
   (b) bucket
   (c) spoon
   (d) measuring cup

2. swimming pool

3. Yes

4. (a) A
   (b) C
   (c) B

5. B

6. A

7. 6
   3
   cup

Activity

Tell your student that *capacity* is the total amount of liquid a container can hold. Show him two of the containers with similar capacity and ask him which holds more water, or has the greatest capacity. Use containers that he cannot just look at and be sure they have the same capacity – one might be taller and thinner than the other. Ask him how he can be sure of his answer. Allow him to demonstrate his ideas.

Use one large container and two much smaller ones, of different capacity, such as a cup and a mug. Ask your student to determine whether the cup or the mug has the greater capacity. Then have her fill up the large container using the cup first, emptying it, and then using the mug. Record the number of cups and mugs that are required, such as 8 cups and a bit (or about 8 cups) and almost 5 mugs. Then discuss why it takes more cups than mugs to fill the container.

Discussion

Page 102

Tasks 1-3, pp. 102-105

In Task 3(a), the mug holds more, since fewer are needed to fill the bowl In (b), the cup will still hold the least, since it holds less than the mug (from (a)) and less than the bottle, since the mug holds less than the bottle.

Workbook

Exercise 1, pp. 163-165

Reinforcement

Extra *Practice*, Unit 10, Exercise 1, pp. 93-94

Tests

Tests, Unit 10, 1A and 1B, pp. 123-128

Chapter 2 – Measuring Capacity

Objectives

- Measure capacity with non-standard units.

Material

- Several containers of different capacity
- One larger container with see-through sides
- Paper cup

Notes

In this section, students will measure capacity using non-standard units, such as the number of glasses of water the container can hold. They will also compare capacity by pouring the full containers into a larger container and marking the height of the water.

In *Primary Mathematics* 2 students will use standard units to measure capacity – liters, gallons, quarts, pints, and cups.

(1) Measure capacity

Textbook

Page 106

   5
   3
   jug
   bottle

Task 1, p. 107

1. (a) C
   (b) D
   (c) A
   (d) B
   (e) C, A, B, D

Workbook

Exercise 2, pp. 166-167

1. 4
  6
  8

2. (a) 6
   (b) 4
   (c) pitcher

3. (a) C
   (b) D
   (c) A

Activity

Use two containers. Mark a level on one with a marker pen and have your student fill it with water to that level. Then have him pour the water into the other container and mark the level. Point out that the amount of water does not change, even though the level marked is different. Discuss why the level is different (e.g. one container is wider). Allow him to experiment with other containers.

Mark a cup near the top, and have your student fill the cup to the mark and then pour it into a larger container. In this manner, measure the capacity of two or more containers by seeing how many cups or glasses of water are needed to fill each one. Record the results and have him line up the containers in order of capacity.

Tell your student that the cup is one unit of capacity. Fill a large transparent container, such as a liter pop bottle, with water, and then help your student pour the water to fill the cup to the mark. Mark the new level. Have her continue transferring water to the cup and marking the new level. Tell your student that each mark is one unit.

Have your student use the marked container to measure the capacity of several smaller containers, (as in task 1, textbook p. 107).

Discussion

Page 106

Task 1, p. 107

Workbook

Exercise 2, pp. 166-167

Reinforcement

Extra Practice, Unit 10, Exercise 2, pp. 95-96

Tests

Tests, Unit 8, 2A and 2B, pp. 129-132

Review

Tests

Tests, Cumulative Test Units 1-10, A and B, pp. 133-142

Workbook

Review 6, pp. 168-172

1.	12	11	2.	13	19
	16	17		11	12
				20	15
				16	14

3. (a) 17   (b) 7
4. $14 - 6 = 4 + 4$
   $12 - 3 = 7 + 2$
   $15 - 8 = 2 + 5$
   $14 - 9 = 1 + 4$
5. the last triangle
6. the rectangle
7. globe, frisbee
8. (a) R; Q   (b) C; B; C
7. $19 - 5 = 14$; 14
8. $12 - 8 = 4$; 4
9. $8 + 7 = 15$; 15

Review 7, pp. 173-176

1. (a) twelve   (b) twenty
2. (a) 3 watermelons colored   (b) 3rd pineapple colored
3. 3, 6, 7, 11, 14
4. (a) 12   (b) 3
5. (a) 13   (b) 5
6. (a) 10   (b) 8
   (c) 16   (d) 13
   (e) 14   (f) 9
7. (a) A; E   (b) C   (c) B
8. the last figure
9. the first and last figure
10. the third container
11. $10 - 1 = 9$; 9
11. $14 - 7 = 7$; 7
12. $16 - 9 = 7$; 7

Mental Math 1	Mental Math 2	Mental Math 3
5 + 0 = 5	4 + 2 = 6	6 + 1 = 7
1 + 4 = 5	6 + 3 = 9	2 + 4 = 6
5 + 1 = 6	3 + 5 = 8	1 + 8 = 9
7 + 1 = 8	1 + 9 = 10	7 + 3 = 10
3 + 2 = 5	2 + 5 = 7	2 + 8 = 10
1 + 2 = 3	7 + 1 = 8	3 + 5 = 8
0 + 4 = 4	5 + 3 = 8	9 + 1 = 10
2 + 3 = 5	2 + 7 = 9	5 + 2 = 7
1 + 5 = 6	3 + 4 = 7	3 + 6 = 9
2 + 2 = 4	3 + 7 = 10	7 + 2 = 9
9 + 0 = 9	1 + 6 = 7	4 + 3 = 7
4 + 1 = 5	6 + 2 = 8	1 + 7 = 8
1 + 6 = 7	3 + 3 = 6	2 + 3 = 5
8 + 1 = 9	8 + 2 = 10	6 + 3 = 9
2 + 2 = 4	3 + 2 = 5	2 + 6 = 8
0 + 8 = 8	3 + 6 = 9	3 + 3 = 6
6 + 1 = 7	4 + 3 = 7	2 + 7 = 9
1 + 7 = 8	7 + 2 = 9	5 + 3 = 8
9 + 0 = 9	8 + 1 = 9	3 + 4 = 7
2 + 1 = 3	5 + 2 = 7	8 + 2 = 10

Mental Math 4	Mental Math 5	Mental Math 6
1 + 1 = 2	8 + 2 = 10	5 − 1 = 4
3 + 2 = 5	4 + 5 = 9	4 − 3 = 1
6 + 2 = 8	1 + 1 = 2	3 − 3 = 0
3 + 3 = 6	3 + 4 = 7	4 − 1 = 3
3 + 7 = 10	5 + 5 = 10	2 − 0 = 2
2 + 8 = 10	6 + 4 = 10	2 − 1 = 1
4 + 6 = 10	0 + 6 = 6	1 − 0 = 1
5 + 4 = 9	8 + 2 = 10	5 − 4 = 1
3 + 4 = 7	4 + 3 = 7	4 − 2 = 2
2 + 2 = 4	2 + 3 = 5	2 − 2 = 0
0 + 8 = 8	7 + 3 = 10	5 − 5 = 0
9 + 1 = 10	4 + 4 = 8	3 − 0 = 3
4 + 5 = 9	3 + 5 = 8	4 − 4 = 0
5 + 3 = 8	1 + 9 = 10	5 − 3 = 2
4 + 4 = 8	5 + 4 = 9	3 − 2 = 1
4 + 6 = 10	2 + 8 = 10	5 − 0 = 5
6 + 3 = 9	3 + 6 = 9	3 − 1 = 2
5 + 5 = 10	2 + 7 = 9	1 − 1 = 0
7 + 0 = 7	3 + 3 = 6	4 − 0 = 4
7 + 2 = 9	3 + 6 = 9	5 − 2 = 3

Mental Math 7	Mental Math 8	Mental Math 9
6 − 1 = 5	8 − 6 = 2	7 − 4 = 3
5 − 2 = 3	10 − 4 = 6	8 − 4 = 4
7 − 1 = 6	6 − 4 = 2	3 − 2 = 1
5 − 3 = 2	10 − 1 = 9	10 − 7 = 3
9 − 1 = 8	10 − 9 = 1	10 − 8 = 2
10 − 2 = 8	8 − 7 = 1	8 − 5 = 3
8 − 2 = 6	10 − 5 = 5	9 − 4 = 5
10 − 1 = 9	9 − 7 = 2	6 − 6 = 0
9 − 2 = 7	7 − 6 = 1	9 − 6 = 3
6 − 2 = 4	10 − 2 = 8	7 − 4 = 3
7 − 3 = 4	9 − 8 = 1	6 − 3 = 3
8 − 3 = 5	10 − 9 = 1	9 − 5 = 4
7 − 2 = 5	6 − 5 = 1	8 − 5 = 3
4 − 3 = 1	10 − 3 = 7	5 − 2 = 3
9 − 3 = 6	5 − 4 = 1	10 − 4 = 6
7 − 3 = 4	10 − 10 = 0	9 − 6 = 3
6 − 3 = 3	10 − 6 = 4	10 − 5 = 5
8 − 1 = 7	5 − 3 = 2	9 − 5 = 4
8 − 3 = 5	10 − 7 = 3	10 − 6 = 4
10 − 3 = 7	4 − 3 = 1	7 − 3 = 4

Mental Math 10	Mental Math 11
4 + 5 = 9	10 − 1 = 4 + 5
7 − 4 = 3	10 − 4 = 9 − 3
9 − 3 = 6	6 − 4 = 10 − 8
5 + 2 = 7	10 − 3 = 3 + 4
8 − 4 = 4	10 − 9 = 1 + 0
6 + 2 = 8	8 − 7 = 9 − 8
10 − 7 = 3	10 − 5 = 3 + 2
9 − 4 = 5	9 − 2 = 4 + 3
4 + 5 = 9	7 − 5 = 10 − 8
8 − 5 = 3	10 − 2 = 5 + 3
3 + 4 = 7	9 − 8 = 5 − 4
7 − 5 = 2	4 + 1 = 9 − 4
9 − 5 = 4	8 − 5 = 10 − 7
4 + 4 = 8	8 − 3 = 2 + 3
9 − 6 = 3	1 + 1 = 7 − 5
10 − 4 = 6	10 − 4 = 9 − 3
6 − 3 = 3	9 − 0 = 6 + 3
8 + 2 = 10	5 − 3 = 8 − 6
8 − 3 = 5	10 − 7 = 2 + 1
7 + 3 = 10	4 − 1 = 7 − 4

Mental Math 12	Mental Math 13
9 + 1 = 10 + **0**	7 + 4 = 10 + **1**
9 + 2 = 10 + **1**	4 + 8 = 10 + **2**
7 + 3 = 10 + **0**	9 + 7 = 10 + **6**
8 + 6 = 10 + **4**	7 + 5 = 10 + **2**
9 + 9 = 10 + **8**	8 + 5 = 10 + **3**
6 + 5 = 10 + **1**	7 + 9 = 10 + **6**
8 + 3 = 10 + **1**	6 + 4 = 10 + **0**
5 + 5 = 10 + **0**	9 + 3 = 10 + **2**
9 + 5 = 10 + **4**	8 + 2 = 10 + **0**
8 + 4 = 10 + **2**	7 + 6 = 10 + **3**
6 + 6 = 10 + **2**	8 + 4 = 10 + **2**
5 + 7 = 10 + **2**	6 + 7 = 10 + **3**
3 + 8 = 10 + **1**	2 + 9 = 10 + **1**
2 + 9 = 10 + **1**	7 + 8 = 10 + **5**
4 + 7 = 10 + **1**	9 + 6 = 10 + **5**
7 + 7 = 10 + **4**	6 + 8 = 10 + **4**
5 + 8 = 10 + **3**	5 + 6 = 10 + **1**
9 + 6 = 10 + **5**	8 + 8 = 10 + **6**
3 + 9 = 10 + **2**	6 + 9 = 10 + **5**
8 + 7 = 10 + **5**	5 + 9 = 10 + **4**

Mental Math 14	Mental Math 15	Mental Math 16
8 + 3 = **11**	3 + 4 = **7**	7 − 4 = **3**
9 + 6 = **15**	13 + 4 = **17**	17 − 4 = **13**
5 + 7 = **12**	4 + 13 = **17**	6 − 3 = **3**
9 + 2 = **11**	8 + 1 = **9**	16 − 3 = **13**
6 + 8 = **14**	18 + 1 = **19**	15 − 2 =**13**
8 + 8 = **16**	8 + 11 = **19**	18 − 5 = **13**
7 + 6 = **13**	5 + 4 = **9**	19 − 6 = **13**
3 + 9 = **12**	15 + 4 = **19**	19 − 2 = **17**
8 + 5 = **13**	5 + 14 = **19**	18 − 4 = **14**
7 + 9 = **16**	17 + 2 = **19**	13 − 2 = **11**
5 + 6 = **11**	14 + 4 = **18**	18 − 5 = **13**
9 + 9 = **18**	5 + 12 = **17**	19 − 4 = **15**
4 + 8 = **12**	11 + 7 = **18**	16 − 6 = **10**
8 + 9 = **17**	3 + 10 = **13**	17 − 3 = **14**
6 + 6 = **12**	16 + 2 = **18**	18 − 6 = **12**
7 + 7 = **14**	12 + 3 = **15**	17 − 4 = **13**
9 + 4 = **13**	13 + 5 = **18**	19 − 5 = **14**
4 + 7 = **11**	2 + 12 = **14**	16 − 4 = **12**
8 + 7 = **15**	1 + 12 = **13**	20 − 6 = **14**
5 + 9 = **14**	10 + 8 = **18**	13 − 3 = **10**

Mental Math 17	Mental Math 18
12 − 10 = 0 + **2**	11 − 9 = **2**
12 − 9 = 1 + **2**	12 − 4 = **8**
12 − 8 = 2 + **2**	14 − 6 = **8**
12 − 6 = 4 + **2**	13 − 7 = **6**
12 − 9 = 1 + 2	11 − 5 = **6**
12 − 8 = **2** + 2	14 − 9 = **5**
14 − 8 = 2 + **4**	12 − 9 = **3**
14 − 6 = 4 + **4**	13 − 5 = **8**
13 − 7 = 3 + **3**	15 − 8 = **7**
12 − 5 = **5** + 2	12 − 8 = **4**
17 − 9 = **1** + 7	16 − 7 = **9**
15 − 7 = **3** + 5	11 − 6 = **5**
16 − 9 = **1** + 6	13 − 4 = **9**
14 − 8 = **2** + 4	17 − 9 = **8**
11 − 4 = **6** + 1	14 − 5 = **9**
12 − 3 = **7** + 2	11 − 2 = **9**
13 − 6 = **4** + 3	20 − 7 = **13**
20 − 9 = 10 + **1**	12 − 5 = **7**
20 − 6 = 10 + **4**	16 − 6 = **10**
20 − 3 = **10** + 7	20 − 4 = **16**

Mental Math 19	Mental Math 20	Mental Math 21
12 − 7 = **5**	9 + 2 = **11**	12 − 4 = **8**
15 − 6 = **9**	15 − 3 = **12**	16 − 8 = **8**
16 − 9 = **7**	11 − 3 = **8**	9 + 9 = **18**
11 − 8 = **3**	13 + 2 = **15**	11 − 2 = **9**
13 − 9 = **4**	20 − 2 = **18**	8 + 7 = **15**
20 − 4 = **16**	12 + 3 = **15**	13 − 8 = **5**
12 − 3 = **9**	13 − 2 = **11**	5 + 9 = **14**
16 − 8 = **8**	16 − 1 = **15**	18 − 9 = **9**
11 − 7 = **4**	8 + 3 = **11**	12 − 8 = **4**
18 − 8 = **10**	11 − 2 = **9**	6 + 7 = **13**
13 − 8 = **5**	14 + 1 = **15**	8 − 3 = **5**
14 − 7 = **7**	12 − 3 = **9**	13 − 4 = **9**
12 − 6 = **6**	14 − 3 = **11**	5 + 5 = **10**
17 − 8 = **9**	9 + 3 = **12**	14 − 7 = **7**
11 − 4 = **7**	19 − 2 = **17**	12 − 9 = **3**
14 − 8 = **6**	12 − 2 = **10**	4 + 8 = **12**
15 − 7 = **8**	20 − 1 = **19**	15 − 7 = **8**
11 − 3 = **8**	16 + 2 = **18**	19 − 5 = **14**
13 − 6 = **7**	20 − 3 = **17**	10 − 6 = **4**
20 − 5 = **15**	17 + 3 = **20**	11 − 6 = **5**

Mental Math 22	Mental Math 23	Mental Math 24
11 − 3 = **8**	13 − 6 = **7**	15 − 6 = **9**
14 − 8 = **6**	3 + 9 = **12**	8 + 5 = **13**
9 + 4 = **13**	14 − 5 = **9**	7 + 7 = **14**
8 + 8 = **16**	7 + 9 = **16**	11 − 9 = **2**
12 − 3 = **9**	2 + 8 = **10**	14 − 6 = **8**
9 + 8 = **17**	11 − 4 = **7**	16 − 4 = **12**
13 − 5 = **8**	6 + 6 = **12**	9 + 6 = **15**
16 − 9 = **7**	14 − 9 = **5**	12 − 5 = **7**
14 + 3 = **17**	10 − 8 = **2**	9 + 2 = **11**
17 − 5 = **12**	6 + 8 = **14**	17 − 9 = **8**
11 − 7 = **4**	17 − 8 = **9**	3 + 3 = **6**
6 + 2 = **8**	20 − 7 = **13**	11 − 5 = **6**
15 − 8 = **7**	11 − 8 = **3**	5 + 6 = **11**
8 + 3 = **11**	7 + 4 = **11**	3 + 7 = **10**
13 − 9 = **4**	15 − 5 = **10**	15 + 5 = **20**
5 + 7 = **12**	12 − 6 = **6**	13 − 7 = **6**
9 − 3 = **6**	15 − 9 = **6**	20 + 0 = **20**
7 + 2 = **9**	5 + 4 = **9**	13 + 4 = **17**
12 + 8 = **20**	11 + 9 = **20**	20 − 2 = **18**
12 − 7 = **5**	12 − 2 = **10**	16 − 7 = **9**

Mental Math 1	Mental Math 2	Mental Math 3
5 + 0 = _____	4 + 2 = _____	6 + 1 = _____
1 + 4 = _____	6 + 3 = _____	2 + 4 = _____
5 + 1 = _____	3 + 5 = _____	1 + 8 = _____
7 + 1 = _____	1 + 9 = _____	7 + 3 = _____
3 + 2 = _____	2 + 5 = _____	2 + 8 = _____
1 + 2 = _____	7 + 1 = _____	3 + 5 = _____
0 + 4 = _____	5 + 3 = _____	9 + 1 = _____
2 + 3 = _____	2 + 7 = _____	5 + 2 = _____
1 + 5 = _____	3 + 4 = _____	3 + 6 = _____
2 + 2 = _____	3 + 7 = _____	7 + 2 = _____
9 + 0 = _____	1 + 6 = _____	4 + 3 = _____
4 + 1 = _____	6 + 2 = _____	1 + 7 = _____
1 + 6 = _____	3 + 3 = _____	2 + 3 = _____
8 + 1 = _____	8 + 2 = _____	6 + 3 = _____
2 + 2 = _____	3 + 2 = _____	2 + 6 = _____
0 + 8 = _____	3 + 6 = _____	3 + 3 = _____
6 + 1 = _____	4 + 3 = _____	2 + 7 = _____
1 + 7 = _____	7 + 2 = _____	5 + 3 = _____
9 + 0 = _____	8 + 1 = _____	3 + 4 = _____
2 + 1 = _____	5 + 2 = _____	8 + 2 = _____

Mental Math 4	Mental Math 5	Mental Math 6
1 + 1 = _____	8 + 2 = _____	5 − 1 = _____
3 + 2 = _____	4 + 5 = _____	4 − 3 = _____
6 + 2 = _____	1 + 1 = _____	3 − 3 = _____
3 + 3 = _____	3 + 4 = _____	4 − 1 = _____
3 + 7 = _____	5 + 5 = _____	2 − 0 = _____
2 + 8 = _____	6 + 4 = _____	2 − 1 = _____
4 + 6 = _____	0 + 6 = _____	1 − 0 = _____
5 + 4 = _____	8 + 2 = _____	5 − 4 = _____
3 + 4 = _____	4 + 3 = _____	4 − 2 = _____
2 + 2 = _____	2 + 3 = _____	2 − 2 = _____
0 + 8 = _____	7 + 3 = _____	5 − 5 = _____
9 + 1 = _____	4 + 4 = _____	3 − 0 = _____
4 + 5 = _____	3 + 5 = _____	4 − 4 = _____
5 + 3 = _____	1 + 9 = _____	5 − 3 = _____
4 + 4 = _____	5 + 4 = _____	3 − 2 = _____
4 + 6 = _____	2 + 8 = _____	5 − 0 = _____
6 + 3 = _____	3 + 6 = _____	3 − 1 = _____
5 + 5 = _____	2 + 7 = _____	1 − 1 = _____
7 + 0 = _____	3 + 3 = _____	4 − 0 = _____
7 + 2 = _____	3 + 6 = _____	5 − 2 = _____

Mental Math 7	Mental Math 8	Mental Math 9
6 − 1 = _____	8 − 6 = _____	7 − 4 = _____
5 − 2 = _____	10 − 4 = _____	8 − 4 = _____
7 − 1 = _____	6 − 4 = _____	3 − 2 = _____
5 − 3 = _____	10 − 1 = _____	10 − 7 = _____
9 − 1 = _____	10 − 9 = _____	10 − 8 = _____
10 − 2 = _____	8 − 7 = _____	8 − 5 = _____
8 − 2 = _____	10 − 5 = _____	9 − 4 = _____
10 − 1 = _____	9 − 7 = _____	6 − 6 = _____
9 − 2 = _____	7 − 6 = _____	9 − 6 = _____
6 − 2 = _____	10 − 2 = _____	7 − 4 = _____
7 − 3 = _____	9 − 8 = _____	6 − 3 = _____
8 − 3 = _____	10 − 9 = _____	9 − 5 = _____
7 − 2 = _____	6 − 5 = _____	8 − 5 = _____
4 − 3 = _____	10 − 3 = _____	5 − 2 = _____
9 − 3 = _____	5 − 4 = _____	10 − 4 = _____
7 − 3 = _____	10 − 10 = _____	9 − 6 = _____
6 − 3 = _____	10 − 6 = _____	10 − 5 = _____
8 − 1 = _____	5 − 3 = _____	9 − 5 = _____
8 − 3 = _____	10 − 7 = _____	10 − 6 = _____
10 − 3 = _____	4 − 3 = _____	7 − 3 = _____

Mental Math 10	Mental Math 11
4 + 5 = _____	10 − 1 = 4 + _____
7 − 4 = _____	10 − 4 = 9 − _____
9 − 3 = _____	6 − 4 = 10 − _____
5 + 2 = _____	10 − 3 = 3 + _____
8 − 4 = _____	10 − 9 = 1 + _____
6 + 2 = _____	8 − 7 = 9 − _____
10 − 7 = _____	10 − 5 = 3 + _____
9 − 4 = _____	9 − 2 = 4 + _____
4 + 5 = _____	7 − 5 = 10 − _____
8 − 5 = _____	10 − 2 = 5 + _____
3 + 4 = _____	9 − 8 = 5 − _____
7 − 5 = _____	4 + 1 = 9 − _____
9 − 5 = _____	8 − 5 = 10 − _____
4 + 4 = _____	8 − 3 = 2 + _____
9 − 6 = _____	1 + 1 = 7 − _____
10 − 4 = _____	10 − 4 = 9 − _____
6 − 3 = _____	9 − 0 = 6 + _____
8 + 2 = _____	5 − 3 = 8 − _____
8 − 3 = _____	10 − 7 = 2 + _____
7 + 3 = _____	4 − 1 = 7 − _____

Mental Math 12	Mental Math 13
9 + 1 = 10 + _____	7 + 4 = 10 + _____
9 + 2 = 10 + _____	4 + 8 = 10 + _____
7 + 3 = 10 + _____	9 + 7 = 10 + _____
8 + 6 = 10 + _____	7 + 5 = 10 + _____
9 + 9 = 10 + _____	8 + 5 = 10 + _____
6 + 5 = 10 + _____	7 + 9 = 10 + _____
8 + 3 = 10 + _____	6 + 4 = 10 + _____
5 + 5 = 10 + _____	9 + 3 = 10 + _____
9 + 5 = 10 + _____	8 + 2 = 10 + _____
8 + 4 = 10 + _____	7 + 6 = 10 + _____
6 + 6 = 10 + _____	8 + 4 = 10 + _____
5 + 7 = 10 + _____	6 + 7 = 10 + _____
3 + 8 = 10 + _____	2 + 9 = 10 + _____
2 + 9 = 10 + _____	7 + 8 = 10 + _____
4 + 7 = 10 + _____	9 + 6 = 10 + _____
7 + 7 = 10 + _____	6 + 8 = 10 + _____
5 + 8 = 10 + _____	5 + 6 = 10 + _____
9 + 6 = 10 + _____	8 + 8 = 10 + _____
3 + 9 = 10 + _____	6 + 9 = 10 + _____
8 + 7 = 10 + _____	5 + 9 = 10 + _____

Mental Math 14	Mental Math 15	Mental Math 16
$8 + 3 =$ _____	$3 + 4 =$ _____	$7 - 4 =$ _____
$9 + 6 =$ _____	$13 + 4 =$ _____	$17 - 4 =$ _____
$5 + 7 =$ _____	$4 + 13 =$ _____	$6 - 3 =$ _____
$9 + 2 =$ _____	$8 + 1 =$ _____	$16 - 3 =$ _____
$6 + 8 =$ _____	$18 + 1 =$ _____	$15 - 2 =$ _____
$8 + 8 =$ _____	$8 + 11 =$ _____	$18 - 5 =$ _____
$7 + 6 =$ _____	$5 + 4 =$ _____	$19 - 6 =$ _____
$3 + 9 =$ _____	$15 + 4 =$ _____	$19 - 2 =$ _____
$8 + 5 =$ _____	$5 + 14 =$ _____	$18 - 4 =$ _____
$7 + 9 =$ _____	$17 + 2 =$ _____	$13 - 2 =$ _____
$5 + 6 =$ _____	$14 + 4 =$ _____	$18 - 5 =$ _____
$9 + 9 =$ _____	$5 + 12 =$ _____	$19 - 4 =$ _____
$4 + 8 =$ _____	$11 + 7 =$ _____	$16 - 6 =$ _____
$8 + 9 =$ _____	$3 + 10 =$ _____	$17 - 3 =$ _____
$6 + 6 =$ _____	$16 + 2 =$ _____	$18 - 6 =$ _____
$7 + 7 =$ _____	$12 + 3 =$ _____	$17 - 4 =$ _____
$9 + 4 =$ _____	$13 + 5 =$ _____	$19 - 5 =$ _____
$4 + 7 =$ _____	$2 + 12 =$ _____	$16 - 4 =$ _____
$8 + 7 =$ _____	$1 + 12 =$ _____	$20 - 6 =$ _____
$5 + 9 =$ _____	$10 + 8 =$ _____	$13 - 3 =$ _____

Mental Math 17	Mental Math 18
$12 - 10 = 0 + \underline{\hspace{2em}}$	$11 - 9 = \underline{\hspace{2em}}$
$12 - 9 = 1 + \underline{\hspace{2em}}$	$12 - 4 = \underline{\hspace{2em}}$
$12 - 8 = 2 + \underline{\hspace{2em}}$	$14 - 6 = \underline{\hspace{2em}}$
$12 - 6 = 4 + \underline{\hspace{2em}}$	$13 - 7 = \underline{\hspace{2em}}$
$12 - 9 = \underline{\hspace{2em}} + 2$	$11 - 5 = \underline{\hspace{2em}}$
$12 - 8 = \underline{\hspace{2em}} + 2$	$14 - 9 = \underline{\hspace{2em}}$
$14 - 8 = 2 + \underline{\hspace{2em}}$	$12 - 9 = \underline{\hspace{2em}}$
$14 - 6 = 4 + \underline{\hspace{2em}}$	$13 - 5 = \underline{\hspace{2em}}$
$13 - 7 = 3 + \underline{\hspace{2em}}$	$15 - 8 = \underline{\hspace{2em}}$
$12 - 5 = \underline{\hspace{2em}} + 2$	$12 - 8 = \underline{\hspace{2em}}$
$17 - 9 = \underline{\hspace{2em}} + 7$	$16 - 7 = \underline{\hspace{2em}}$
$15 - 7 = \underline{\hspace{2em}} + 5$	$11 - 6 = \underline{\hspace{2em}}$
$16 - 9 = \underline{\hspace{2em}} + 6$	$13 - 4 = \underline{\hspace{2em}}$
$14 - 8 = \underline{\hspace{2em}} + 4$	$17 - 9 = \underline{\hspace{2em}}$
$11 - 4 = \underline{\hspace{2em}} + 1$	$14 - 5 = \underline{\hspace{2em}}$
$12 - 3 = \underline{\hspace{2em}} + 2$	$11 - 2 = \underline{\hspace{2em}}$
$13 - 6 = \underline{\hspace{2em}} + 3$	$20 - 7 = \underline{\hspace{2em}}$
$20 - 9 = 10 + \underline{\hspace{2em}}$	$12 - 5 = \underline{\hspace{2em}}$
$20 - 6 = 10 + \underline{\hspace{2em}}$	$16 - 6 = \underline{\hspace{2em}}$
$20 - 3 = \underline{\hspace{2em}} + 7$	$20 - 4 = \underline{\hspace{2em}}$

Mental Math 19	Mental Math 20	Mental Math 21
12 − 7 = _____	9 + 2 = _____	12 − 4 = _____
15 − 6 = _____	15 − 3 = _____	16 − 8 = _____
16 − 9 = _____	11 − 3 = _____	9 + 9 = _____
11 − 8 = _____	13 + 2 = _____	11 − 2 = _____
13 − 9 = _____	20 − 2 = _____	8 + 7 = _____
20 − 4 = _____	12 + 3 = _____	13 − 8 = _____
12 − 3 = _____	13 − 2 = _____	5 + 9 = _____
16 − 8 = _____	16 − 1 = _____	18 − 9 = _____
11 − 7 = _____	8 + 3 = _____	12 − 8 = _____
18 − 8 = _____	11 − 2 = _____	6 + 7 = _____
13 − 8 = _____	14 + 1 = _____	8 − 3 = _____
14 − 7 = _____	12 − 3 = _____	13 − 4 = _____
12 − 6 = _____	14 − 3 = _____	5 + 5 = _____
17 − 8 = _____	9 + 3 = _____	14 − 7 = _____
11 − 4 = _____	19 − 2 = _____	12 − 9 = _____
14 − 8 = _____	12 − 2 = _____	4 + 8 = _____
15 − 7 = _____	20 − 1 = _____	15 − 7 = _____
11 − 3 = _____	16 + 2 = _____	19 − 5 = _____
13 − 6 = _____	20 − 3 = _____	10 − 6 = _____
20 − 5 = _____	17 + 3 = _____	11 − 6 = _____

Mental Math 22	Mental Math 23	Mental Math 24
$11 - 3 =$ _____	$13 - 6 =$ _____	$15 - 6 =$ _____
$14 - 8 =$ _____	$3 + 9 =$ _____	$8 + 5 =$ _____
$9 + 4 =$ _____	$14 - 5 =$ _____	$7 + 7 =$ _____
$8 + 8 =$ _____	$7 + 9 =$ _____	$11 - 9 =$ _____
$12 - 3 =$ _____	$2 + 8 =$ _____	$14 - 6 =$ _____
$9 + 8 =$ _____	$11 - 4 =$ _____	$16 - 4 =$ _____
$13 - 5 =$ _____	$6 + 6 =$ _____	$9 + 6 =$ _____
$16 - 9 =$ _____	$14 - 9 =$ _____	$12 - 5 =$ _____
$14 + 3 =$ _____	$10 - 8 =$ _____	$9 + 2 =$ _____
$17 - 5 =$ _____	$6 + 8 =$ _____	$17 - 9 =$ _____
$11 - 7 =$ _____	$17 - 8 =$ _____	$3 + 3 =$ _____
$6 + 2 =$ _____	$20 - 7 =$ _____	$11 - 5 =$ _____
$15 - 8 =$ _____	$11 - 8 =$ _____	$5 + 6 =$ _____
$8 + 3 =$ _____	$7 + 4 =$ _____	$3 + 7 =$ _____
$13 - 9 =$ _____	$15 - 5 =$ _____	$15 + 5 =$ _____
$5 + 7 =$ _____	$12 - 6 =$ _____	$13 - 7 =$ _____
$9 - 3 =$ _____	$15 - 9 =$ _____	$20 + 0 =$ _____
$7 + 2 =$ _____	$5 + 4 =$ _____	$13 + 4 =$ _____
$12 + 8 =$ _____	$11 + 9 =$ _____	$20 - 2 =$ _____
$12 - 7 =$ _____	$12 - 2 =$ _____	$16 - 7 =$ _____

0	1	2
3	4	5
6	7	8

9	10	11
12	13	14
15	16	17

18 19 20

one two

three four

five six

seven

eight

nine

ten

eleven

twelve

thirteen

fourteen

fifteen

sixteen

seventeen

eighteen

nineteen

twenty

zero

1st	2nd	3rd
4th	5th	6th
7th	8th	9th

1 0 10th

+ − =

+ − =

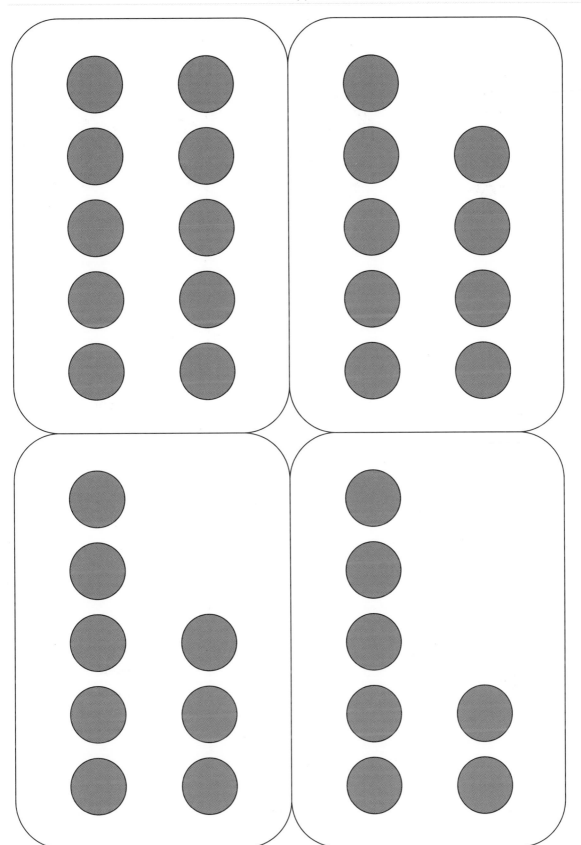

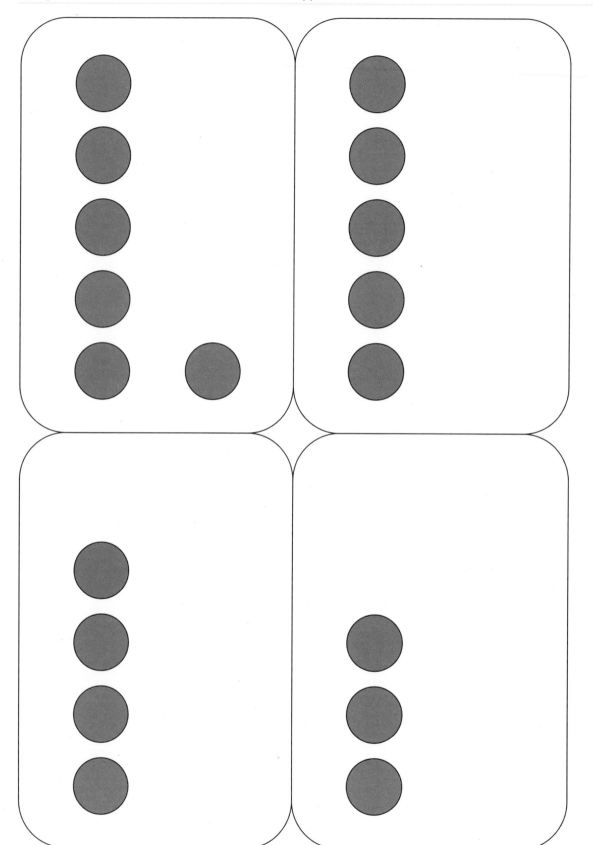

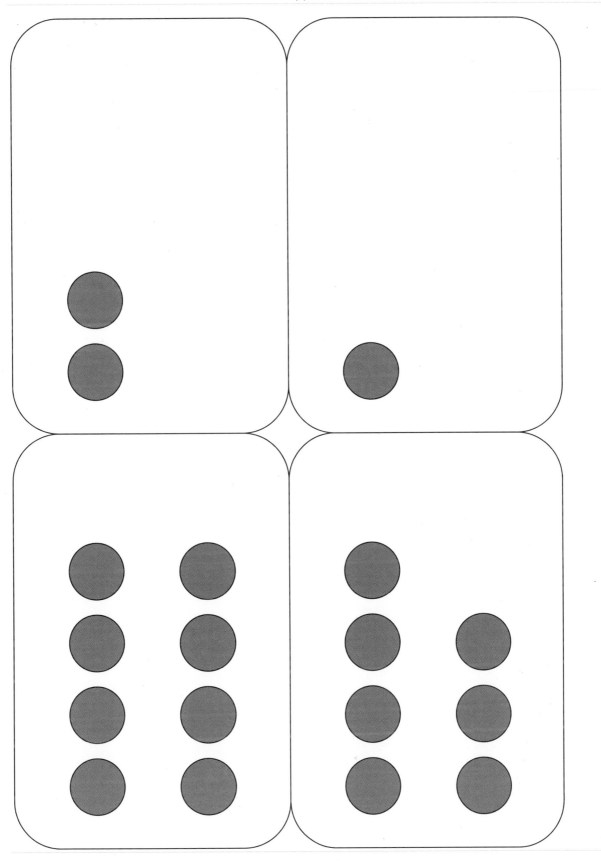

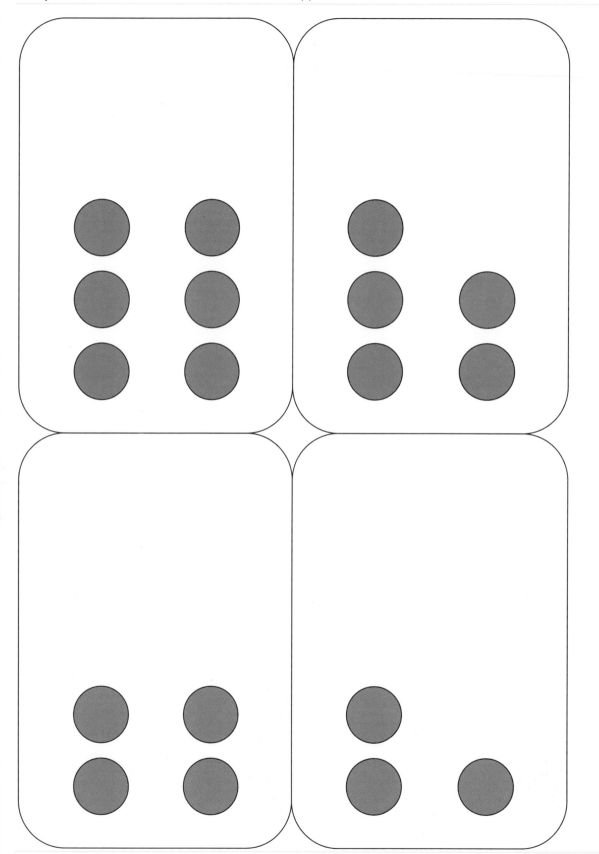

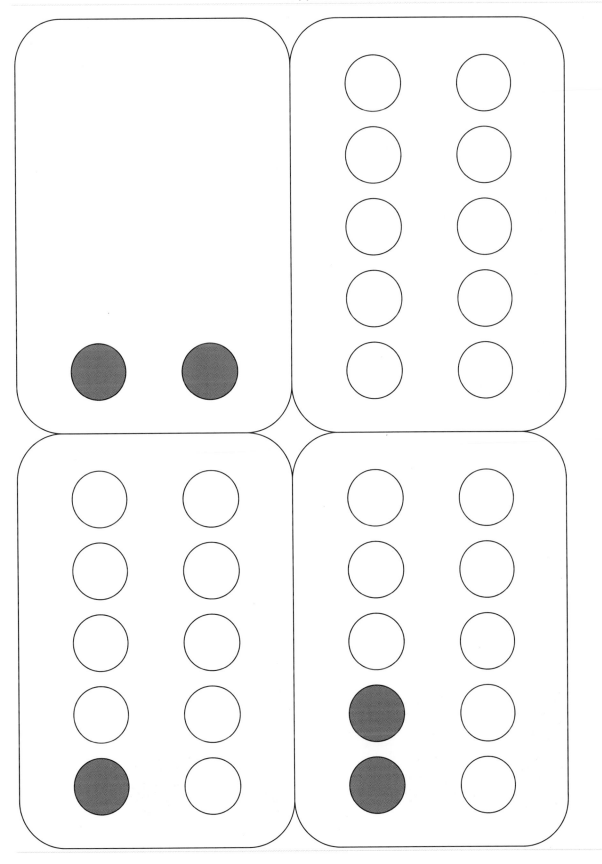

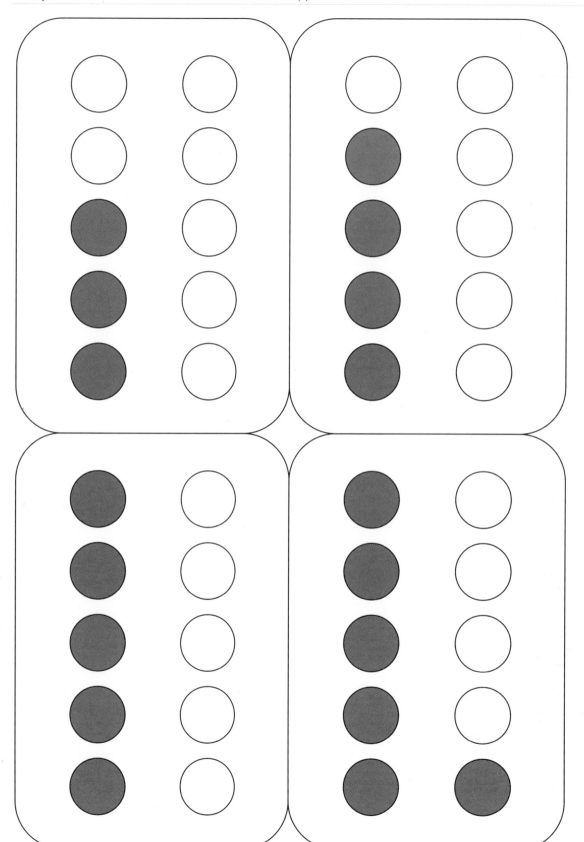

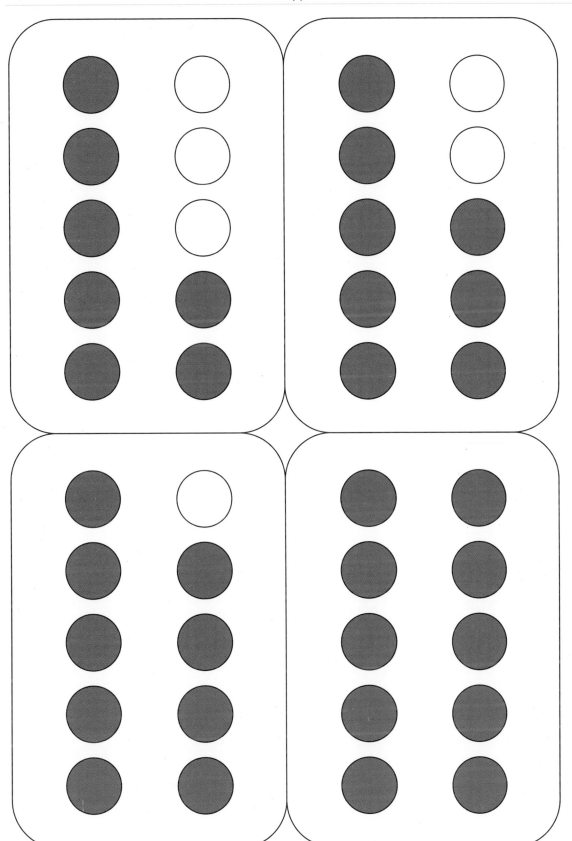

1	2	3	4	5	6	7	8	9	10
11	12	13	14	15	16	17	18	19	20
21	22	23	24	25	26	27	28	29	30
31	32	33	34	35	36	37	38	39	40
41	42	43	44	45	46	47	48	49	50
51	52	53	54	55	56	57	58	59	60
61	62	63	64	65	66	67	68	69	70
71	72	73	74	75	76	77	78	79	80
81	82	83	84	85	86	87	88	89	90
91	92	93	94	95	96	97	98	99	100